Design of Structural Steelwork

Design of Structural Steelwork

Second Edition

PETER KNOWLES, MA, MPhil, CEng, MICE, FIHT
Consulting Engineer

Surrey University Press

Glasgow and London

Published by Surrey University Press
Bishopbriggs, Glasgow G64 2NZ and
7 Leicester Place, London WC2H 7BP

First published 1987

British Library Cataloguing in Publication Data

Knowles, P.R.
Design of structural steelwork. —2nd ed.
1. Steel, Structural
I. Title
624.1′821 TA684
ISBN 0–903384–59–0

Phototypesetting by Thomson Press (I) Limited, New Delhi

Printed in Great Britain by Bell & Bain (Glasgow) Ltd.

Contents

Preface

Instruction in structural design has always been considered an essential part of the training of a student engineer, though the difficulties of teaching the subject effectively have not always been completely appreciated. An ideal course should combine theoretical instruction and practical application; limitations of time, space and money generally restrict the latter aspect to calculation and drawing with perhaps the construction and testing of models. But much can be done with pencil and paper to inculcate a sound approach to the design of structures, provided the student is made aware of the fundamentals of design method and the specific problems associated with the various structural materials.

The aim of this publication is to present the essential design aspects of one structural material—steel. The book is of an entirely introductory nature, demanding no prior knowledge of the subject, but readers are assumed to have followed (or be following) courses in structural analysis and mechanics of materials in sufficient depth to give them a confident grasp of elementary structural and stress analysis techniques. Although it has been written primarily with undergraduates in mind the book will be of use to young graduates who may be coming across the subject for the first time. For this reason the example calculations conform as far as possible to practical requirements.

The first chapter commences with a brief review of the historical development of the science of iron and steel making and the use of these two materials in structures, followed by a discussion of the important properties of structural steel, and the types of steel products available for structural use.

Design philosophy and stability, outlined in Chapter 2, are followed by a detailed chapter on that most important structural element, the beam. After consideration of local and overall instability the chapter goes on to describe the design of a number of different beam types; rolled sections, compound beams, welded plate girders, gantry girders and composite beams.

Chapter 4 is devoted to elements loaded in tension or compression, with or without bending, considering rolled and built-up members, concrete encasement and concrete filling, and the special problems of angle members.

Connections are the subject of Chapter 5. Detailed treatment of the fundamentals of connection design is given, with emphasis on high-strength friction grip bolting and welding. Finally Chapter 6 introduces some very simple assemblies of elements.

Mere manipulation of code of practice clauses is a poor preparation for a student; he must be awarc of the theoretical background to present and future design practice. Yet codified information needs to be used if comparisons are to be made and some discipline imposed on examples and exercises. In this case the current version of British Standard 5950 has been used as a basis for calculations.

Extracts from BS5950: Part 1:1985 are reproduced by permission of the British Standards Institution. Complete copies can be obtained from BSI at Linford Wood, Milton Keynes, MK14 6LE.

Design is an open-ended subject in which there are no unique solutions. Students often have difficulty in accepting this fact, accustomed as they are to finding the unique correct solution to an analytical problem. They must try to cultivate an attitude of mind which will help them to criticise their solution to design problems from economic and aesthetic points of view in so far as this is possible in a student environment.

Finally, an intelligent interest in the world of engineering is essential. Visits to structures under construction, fabricating shops and steelworks are to be encouraged. At the very least students should read architectural and engineering journals to keep abreast of developments in steel construction. It must always be borne in mind that a textbook such as this one must of necessity always lag behind the most modern practice even though the fundamental ideas which it contains will still be valid.

My particular thanks go to Norman Wootton, BSc, MICE, for his help in checking the example calculations.

PK

Notations and units

The system of notation adopted follows that in British Standard 5950. The major symbols are listed here for reference; others are defined when used in the text.

The units adopted are generally those of the SI system with the important variation that the centimetre (which is not in the SI system) has been retained for the steel section properties, radius of gyration (cm), area (cm^2), modulus (cm^3) and second moment of area (cm^4).

The mass of a cubic metre of steel is 7850 kilograms.

1 metric tonne = 9.81 kilonewtons

A	Area
A_e	Effective area
A_g	Gross area
A_s	Shear area (bolts)
A_t	Tensile stress area (bolts)
A_v	Shear area (sections)
a	Spacing of transverse stiffeners *or* Effective throat size of weld
B	Breadth
b	Outstand *or* Width of panel
b_1	Stiff bearing length
D	Depth of section *or* Diameter of section *or* Diameter of hole
d	Depth of web *or* Nominal diameter of fastener
E	Modulus of elasticity of steel
e	End distance
F_c	Compressive force due to axial load
F_s	Shear force (bolts)
F_t	Tensile force
F_v	Average shear force (sections)
f_c	Compressive stress due to axial load
f_v	Shear stress
G	Shear modulus of steel
H	Warping constant of section
h	Storey height
I_x	Second moment of area about the major axis
I_y	Second moment of area about the minor axis
J	Torsion constant of section

L	Length of span
L_E	Effective length
M	Larger end moment
M_{ax}, M_{ay}	Maximum buckling moment about the major or minor axes in the presence of axial load
M_b	Buckling resistance moment (lateral torsional)
M_{cx}, M_{cy}	Moment capacity of section about the major and minor axes in the absence of axial load
M_E	Elastic critical moment
M_o	Midspan moment on a simply supported span equal to the unrestrained length
M_{rx}, M_{ry}	Reduced moment capacity of the section about the major and minor axes in the presence of axial load
$\bar{M}_x, \bar{M}_y$	Applied moment about the major and minor axes
M_x, M_y	Equivalent uniform moment about the major and minor axes
m	Equivalent uniform moment factor
n	Slenderness correction factor
P_{bb}	Bearing capacity of a bolt
P_{bg}	Bearing capacity of parts connected by friction grip fasteners
P_{bs}	Bearing capacity of parts connected by ordinary bolts
P_{cx}, P_{cy}	Compression resistance about the major and minor axes
P_s	Shear capacity of a bolt
P_{sL}	Slip resistance provided by a friction grip fastener
P_t	Tension capacity of a member or fastener
P_v	Shear capacity of a section
p_b	Bending strength
p_{bb}	Bearing strength of a bolt
p_{bg}	Bearing strength of parts connected by friction grip fasteners
p_{bs}	Bearing strength of parts connected by ordinary bolts
p_c	Compressive strength
p_E	Euler strength
p_s	Shear strength of a bolt
p_t	Tension strength of bolt
p_w	Design strength of a fillet weld
p_y	Design strength of steel
q_b	Basic shear strength of a web panel
q_{cr}	Critical shear strength of web panel
q_e	Elastic critical shear strength of web panel
q_f	Flange dependent shear strength factor
r_x, r_y	Radius of gyration of a member about its major and minor axes
S_x, S_y	Plastic modulus about the major and minor axes
s	Leg length of a fillet weld
T	Thickness of a flange or leg
t	Thickness of a web
U_s	Specified minimum ultimate tensile strength of the steel
u	Buckling parameter of the section
V_b	Shear buckling resistance of stiffened web utilizing tension field action
V_{cr}	Shear buckling resistance of stiffened or unstiffened web without utilizing tension field action
v	Slenderness factor for beam
x	Torsional index of section
Y_s	Specified minimum yield strength of steel
Z_x, Z_y	Elastic modulus about major and minor axes
α_e	Modular ratio
β	Ratio of smaller to larger end moment
γ_f	Overall load factor
γ_o	Ratio M/M_o, i.e. the ratio of the larger end moment to the midspan moment on a simply supported span
δ	Deflection

ε	Constant $\left(\frac{275}{p_y}\right)^{\frac{1}{2}}$
λ	Slenderness, i.e. the effective length divided by the radius of gyration
λ_{cr}	Elastic critical load factor
λ_{LO}	Limiting equivalent slenderness
λ_{LT}	Equivalent slenderness
λ_0	Limiting slenderness
μ	Slip factor

A note on calculations

The example calculations have been laid out in a form similar to that adopted in a design office. Reference is made in the left-hand column of the calculation sheet to the relevant clause of the British Standard which affects the calculation in the centre column. Where a British Standard number is not quoted the reference is to British Standard 5950 (1). The student is urged to carry out all calculations in a methodical manner on prepared calculation sheets; in this way the possibility of error will be reduced and checking facilitated.

In order to make the best use of the example calculations a copy of British Standard 5950 (1) and tables of section properties (2) are necessary.

Further example calculations are to be found in Reference (3).

References

1. British Standard 5950. Structural use of Steelwork in Building: Part 1:1985. Code of Practice for Design in Simple and Continuous Construction: Hot Rolled Sections. Part 2: 1985. Specification for Materials. Fabrication and Erection.
2. Steelwork Design. Guide to BS5950: Part 1:1985. Volume 1. Section Properties. Member Capacities. Constrado, London 1985.
3. Steelwork Design. Guide to BS5950; Part 1:1985. Volume 2. Worked Examples. The Steel Construction Institute, London 1986.

1 Iron and steel

1.1 Production

The basic constituent of structural steel is iron, an element widely and liberally available over the world's surface but with rare exceptions found only in combination with other elements. The main deposits of iron are in the form of ores of various kinds which are distinguished by the amount of metallic iron in the combination and the nature of the other elements present. The most common ores are oxides of iron mixed with earthy materials and chemically adulterated with, for example, sulphur and phosphorus.

Iron products have three main commercial forms; wrought iron, steel and cast iron in ascending order of carbon content. Table 1.1, which gives some physical properties of these three compounds, shows that as the carbon content of the metal increases the melting point is lowered; this fact has considerable importance in the production process.

Modern steelmaking depends for its raw material on iron produced by a blast furnace. Iron ore is charged into the furnace with coke and limestone. A powerful air blast raises the temperature sufficiently to melt the iron, which is run off. The iron at this stage has a high carbon content; steel is obtained from it by removing most of the carbon. In the most modern processes decarburizing is done by blowing oxygen through the molten iron (1).

Table 1.1 Some properties of iron and steel

Material	Typical carbon (%)	Melting point (°C)	Ultimate tensile stress (N/mm^2)
Pure iron	nil	1535	335
Mild steel	up to 0.25	varies*	450
High carbon steel	1.4	varies*	900
Cast iron	5.0	1140	110

*Melting point decreases as carbon content increases

1.2 Mechanical working

By the middle of the nineteenth century the practice of rolling iron sections had been established, iron *rails* being exhibited by the Butterley Company at the Hyde Park Exhibition in 1851. The first wrought-iron I section beams were rolled in 1845 in France under the direction of a French Iron-master, F. Zores. It appears that similar sections were first rolled in England in 1863. Dorman Long were rolling steel beams up to 400 mm deep in 1885 but in the United States a tabular presentation of the section properties of rolled steel shapes had been published in 1873.

1.3 Steel in structure

By the end of the eighteenth century all that was needed to inaugurate an era of building in iron was the courage which all pioneers require. The bridge at Coalbrookdale 1777–81 constructed by Abraham Darby III (1750–91) and the iron framed factories designed by William Strutt (1756–1830) from 1792 onwards appear to mark the beginning of this era. At first only cast iron *columns* were used in building but in 1801 James Watt devised a cast iron beam in the form of an inverted T which could span 4.3 m as a floor beam.

In building construction the centre of pressure to adopt steel framing was located first in the United States. The reasons for this were complex; suffice it to say that in Chicago in the 1880s economic factors, stemming from the need to make the greatest use of expensive land in a cramped city centre, led to the adoption of the tall steel-framed building later known as the skyscraper. High building in traditional masonry construction was limited by the great thickness of material required at lower levels and the consequent heavy load imposed on the foundations. The personal physical problem of climbing stairs had been solved by the invention of the elevator in 1857 (E.G. Otis). A six-storey wrought iron frame, the Cooper Union Building was completed in 1858. All these facts, coupled with an aggressive marketing attitude by the American steel makers, produced a climate in which in 1884 William le Barron Jenney (1832–1907) designed the nine-storey Home Insurance Building, the first skeletal iron and steel frame (2). Major steel frame construction in Great Britain is generally agreed to have commenced with the Ritz Hotel (1904) in London.

A summary of significant dates in both building and bridge construction is given below.

Significant dates in iron and steel structural history

Date	*Details*
1779	Cast iron bridge at Coalbrookdale span 30.0 m
1792	Multi-storey iron-framed mill building at Derby (William Strutt)

1796 Buildwas bridge span 43.0 m
1796 Sunderland bridge span 79.0 m
1801 James Watt cast iron beams to span 4.3 m
1809 Schuylkill bridge span 103 m
1820 Berwick bridge span 150 m
1826 Menai suspension bridge span 177 m
1845 Wrought iron beams rolled in France
1848 Five-storey factory New York (James Bogardus)
1850 Menai tubular rail bridge span 153 m
1857 Otis elevator invented
1853–58 Cooper Union six-storey wrought iron frame
1856 Bessemer steel-making process
1860 Boat store Sheerness, four-storey cast iron frame
1863 Butterley Co rolled wrought iron beams
1865 Siemens Martin open hearth steel-making process
1877 Board of Trade regulations changed to allow steel to be used in bridges
1880 Siemens electric lift invented
1883 Brooklyn Bridge span 486 m
1884 Home Insurance Building Chicago, ten-storey steel frame (W. le Baron Jenney)
1884 Garabit viaduct span 180 m (Eiffel)
1885 Dorman Long opened constructional departments
1887 Hexagonal steel columns used in Birmingham
1889 Eiffel Tower 300 m high
1890 Forth Rail bridge span 521 m
1896 Robinsons, Stockton, first steel frame in England
1904 Ritz Hotel London
1917 Quebec Bridge span 549 m
1931 Bayonne Bridge span 510 m
1932 Sydney Harbour Bridge span 509 m
1937 Golden Gate Bridge span 1280 m
1964 Verrazano Narrows Bridge span 1298 m
1981 Humber Bridge span 1410 m

1.4 Properties of structural steel (3)

To the structural designer, certain properties of steel merit special consideration. As a general introduction to the behaviour of steel under load it is helpful to refer to a tensile stress–strain diagram for an average mild (low carbon) steel. This is shown complete in Figure 1.1 and a portion of it to a larger scale in Figure 1.2. From the diagram the following important characteristics of the material can be deduced:

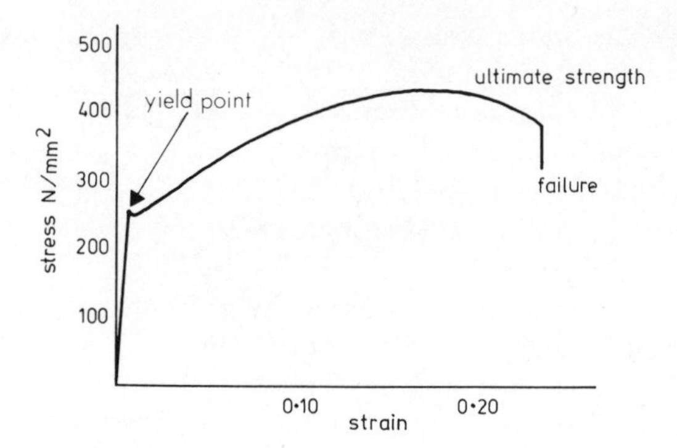

Figure 1.1 Stress–strain curve to failure for typical mild steel

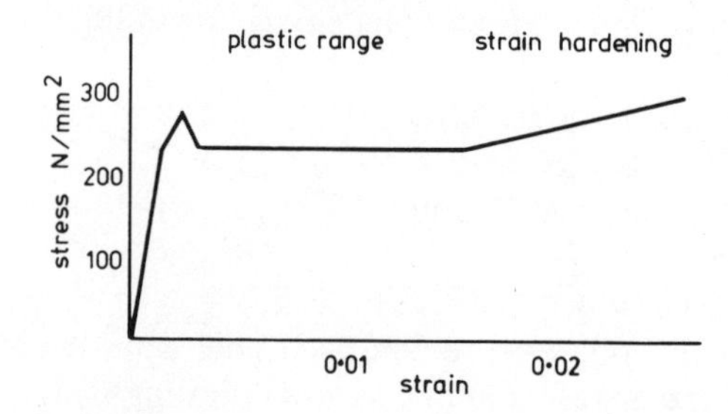

Figure 1.2 Part of the stress–strain curve for typical mild steel

1.4.1 *Elasticity*

Up to a well-defined yield point steel behaves as a perfectly elastic material. Removal of stress at levels below the yield stress causes the material to revert to its unstressed dimensions. Elasticity is also exhibited by higher strength steels which do not have a defined yield point (Figure 1.3). Strictly speaking linear elastic behaviour ceases at a stress level below the yield point known as the proportional limit but this level is difficult to determine and the deviation from straight line behaviour up to yield is very small. The slope of the stress–strain curve in the elastic range defines the modulus of elasticity. For structural steels its value is virtually independent of the steel type and is commonly taken as 205 kN/mm^2.

1.4.2 *Tensile strength*

The applied stress to cause failure is considerably greater than the yield stress; in Figure 1.1 for instance the ultimate stress is nearly twice the yield stress.

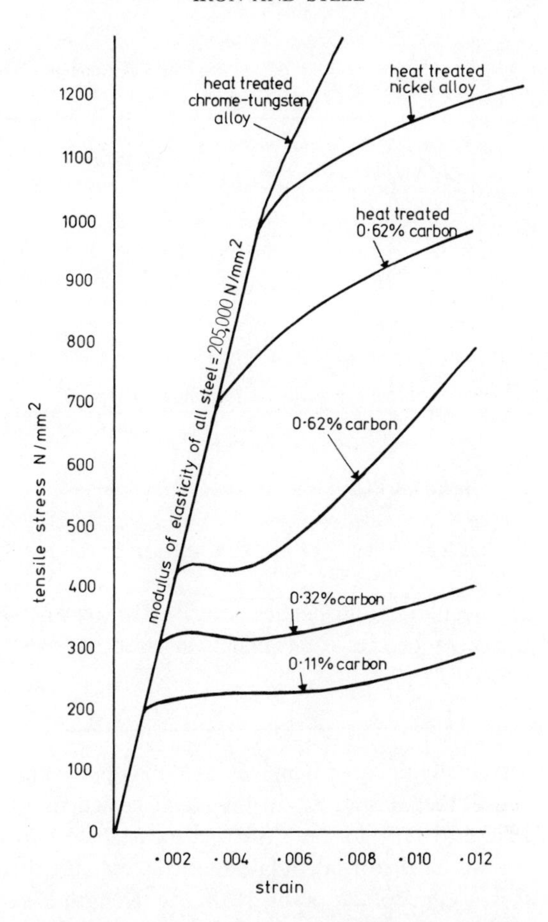

Figure 1.3 Stress–strain curves for higher carbon steels

1.4.3 *Ductility*

An important property of steel is its ability to undergo large deformations without fracture. The strain to failure may reach 25% in a mild steel, will be less for higher carbon steels and may be drastically curtailed in all steels under circumstances which lead to brittle fracture. From an examination of the stress–strain curve, it will be seen that the elastic strain is a small portion of the total strain possible before fracture occurs.

In order to analyse the behaviour of steel elements which are stressed beyond the elastic limit (yield point) there is a need to simplify the real stress–strain curve for steel. A suitable simplification is to replace the portion of the curve from yield to failure by a horizontal line representing strain at constant stress. The resulting *elastic–plastic* stress–strain diagram is illustrated in Figure 1.4. Because it neglects the region of strain hardening the elastic–

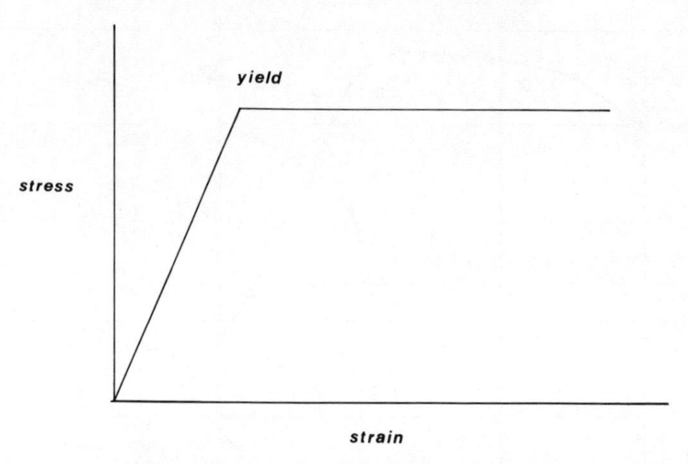

Figure 1.4 Elastic–plastic stress strain diagram

plastic relationship is a conservative approximation to the real strength of the material.

Apart from the mechanical properties described above an awareness of the susceptibility of steel to certain other effects is essential.

1.5 Fire protection

It is ironic that originally uncased iron was used in construction as a fireproof element replacing timber. However, in the steel framed multi-storey blocks built between 1870 and 1900 in the USA the uncased steelwork, though not itself inflammable, was so distorted and weakened after a fire that it was useless in carrying load. The effect of temperature on the strength of steel is shown in Figure 1.5 from which it can be seen that from a temperature of about 300 °C upwards there is a progressive loss of strength. At 600 °C the yield point has fallen to about 0.2 of its value at ordinary temperatures. Clearly where steel framed buildings have contents which, when ignited, can produce high temperatures there is a necessity to provide fire protection for the steelwork.

Much research has been directed to discovering the best method of protecting steel from fire. Building regulations concerned with fire protection have been in existence since the turn of the present century. The severity of fire attack on a structure is determined by the fuel content of the building which is determined by the combustible portions of the structure (joinery etc.) and the furniture, fittings or stored goods in it. Where a building contains incombustible material (e.g. a store containing only metal objects) the fire load is nil. At the other end of the scale a building such as a paint store may contain large quantities of highly inflammable material. Between the two there will be grades of fire load many of which will be insufficient to heat steel above the danger point (4).

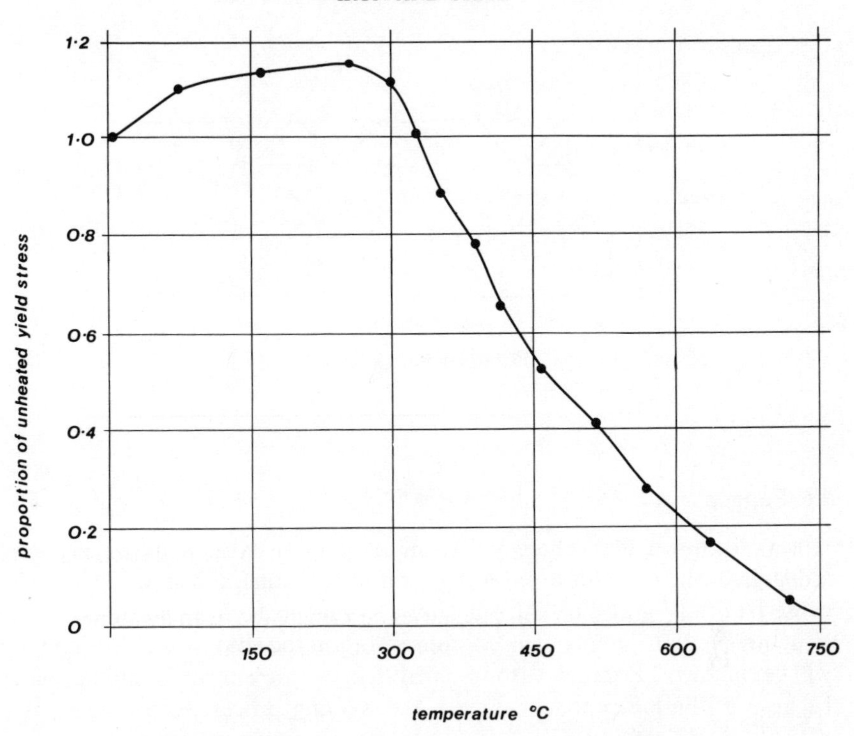

Figure 1.5 Strength–temperature relationship for mild steel

The early forms of fire protection involved the use of a heavy encasement in concrete or brickwork, and concrete is still traditionally thought of as 'fire protection'. However, concrete has limitations, notably in its high weight penalty and also in the work involved in placing it. In addition the use of concrete is restricted to beams and columns; it cannot be used for complex members such as lattice girders, space frames and roof trusses. The weight of concrete casing may add 10% to the total load on the frame, with consequent increase in foundation costs.

For this reason the use of lightweight encasures has become increasingly common. Apart from the reduction in weight they are much more quickly applied, do not need complicated formwork, and are often in dry sheet form. The principal lightweight materials are vermiculite, gypsum and perlite. These may be made up into sheet form; applied as wet plaster on metal lathing; or even sprayed directly on to the structural steelwork. The use of these materials transfers fire protection from the structural to the finishing trade.

Recent innovations in fire protection include hollow columns through which cooling water is circulated, intumescent paint which froths when heated, providing an insulating layer to the steel underneath, and columns placed outside the building away from sources of heat (5).

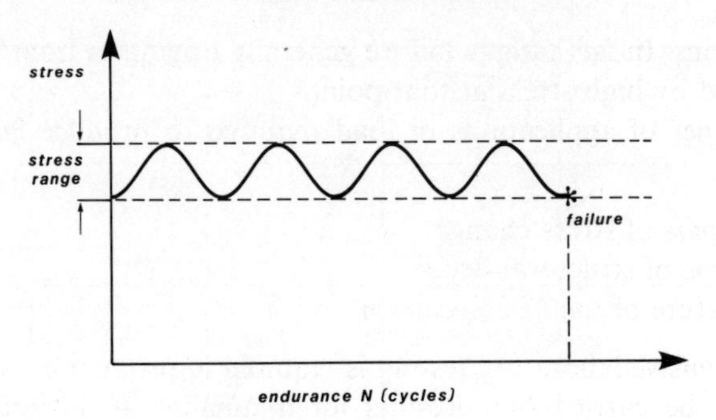

Figure 1.6 Fatigue failure

1.6 Fatigue

It is well known that whereas a structure may sustain an unvarying load indefinitely the effect of a pulsating load of the same maximum value may cause failure (Figure 1.6). Such a failure is said to be due to fatigue. If the structure contains points of stress concentration the effect of a pulsating load will be enhanced. For this reason welded structures are prone to fatigue failure because of the inherent tendency in the welding process to produce stress

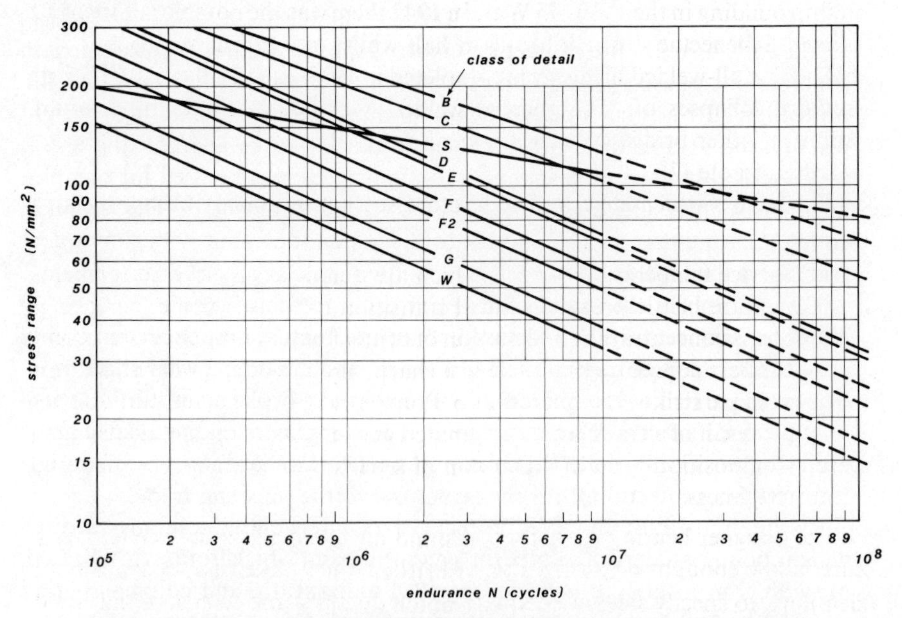

Figure 1.7 Stress range–endurance curves for various classes of steelwork detail

concentrations. In fact fatigue failure generally originates from a very small crack caused by high stress at that point.

The number of applications of load required to produce fatigue failure depends on:

(a) the *range* of stress change
(b) the *type* of structural detail
(c) the nature of the *load spectrum.*

Comprehensive laboratory testing is required to produce data from which design may be carried out. Results for a number of different classes of steelwork detail are shown in Figure 1.7, which can be used to predict the life (number of cycles to failure) for an element subjected to a given stress range (6).

1.7 Brittle fracture

Steel normally behaves as a ductile material having an elongation to failure of about 20% of original length. However, in certain fairly well defined circumstances, it can fail suddenly in a brittle fashion with virtually no deformation and at low stress. The history of brittle failure extends back almost to the Bessemer process in 1856.

Although there are cases of riveted structures failing, it is welded structures which are particularly liable to catastrophic collapse because of the continuous path provided for the fracture. The first all-welded ship was constructed in Great Britain in 1921 but welding really came into prominence in ship building in the 1939–45 War. In 1942 there was the notable failure of T2 tanker 'Schenectady' which broke in half whilst in a fitting out dock.

The first all-welded bridge was completed in 1932. In 1938 there were brittle fracture collapses of a number of welded Vierendeel bridges in Belgium. Amongst other bridge collapses that of Kings Bridge Melbourne in 1962 is worthy of note (7).

The factors which affect brittle fracture susceptibility may be summarised as:

(a) Service temperature—steels which are ductile at an elevated temperature are brittle below a critical transition temperature.
(b) Stress concentration—initiation of brittle fracture may occur at a point of stress concentration such as a sharp corner detail, a weld crack or a weld arc strike. The welded ship 'Ponagansett' broke up in still water as the result of a crack which originated at a tack weld holding a cable clip.
(c) Composition—the composition of a particular type of steel affects its toughness.

The designer bearing these facts in mind must, where the service temperature is low enough to cause a risk of brittle failure, take care to avoid poor detailing, to specify steel of adequate notch ductility and to ensure that weld specification and inspection is correct (8).

Table 1.2 Mechanical properties and carbon content for sections other than hollow sections

Grade	Maximum carbon content (%)	Tensile strength (N/mm^2)	Minimum yield strength for thickness t in mm (N/mm^2)				Minimum elongation on gauge length of 200 mm (%)	Minimum V-notch value Temp (°C)	Charpy impact Energy (J)	Proposed new grade
			$t \leqslant 16$	$16 \leqslant 40$	$49 \leqslant 63$	$63 \leqslant 100$				
43A	0.25	430	275	265	255	245	20	—	—	275A
43D	0.18	430	275	265	255	245	20	−20	27	275D
50B	0.20	490	355	345	340	325	18	—	—	355B
50D	0.18	490	355	345	340	325	18	−20	27	355D
			$t \leqslant 16$	$16 \leqslant 25$	$25 \leqslant 40$			—	—	
55C	0.22	550	450	430	415	—	17	0	27	450C
			$t \leqslant 12$	$12 \leqslant 25$	$25 \leqslant 40$	$40 \leqslant 50$		—	—	
WR50B	0.19	480	345	345	345	340	19	−15	27	WR345B

1.8 Corrosion

Unprotected steel can be severely affected by atmospheric conditions causing rusting and other types of surface degradation. Painting has traditionally been the main type of anti-corrosion treatment adopted and the cost of maintaining the paint film is a significant factor in the total expenditure account for a steel structure. However, improved paint treatments are now available which can give, in reasonable atmospheres, a maintenance free life of the order of 20 years. In addition techniques of metal coating such as galvanizing or zinc spraying provide very good protection (9).

It is now possible to obtain a steel which in normal atmospheres does not require any surface protection. A thin film of oxide forms on its surface, but, unlike the rust on ordinary steel, the oxide does not flake away exposing a fresh surface to corrosion. Instead the initial film adheres tightly to the steel inhibiting further corrosion. Structures with exposed steelwork made from this special steel have been standing successfully for a number of years. The steels have been given the general name of 'weathering steels'.

1.9 Structural steels

In Great Britain structural steel is available commercially in three basic grades, 43, 50 and 55, the numbers representing approximately the minimum tensile strength of each grade in $N/mm^2 \times 10^{-1}$. Within each grade are a number of lettered sub-classes representing steels of increasing notch ductility measured by Charpy impact test and hence increasing resistance to brittle fracture. Grade 50 steels are also available in weather resistant (WR) types. Some properties of these steels are shown in Table 1.2. It may be seen from the table (which contains only selected steels from each grade) that the guaranteed minimum yield stress reduces as the metal thickness is increased. All the steels are weldable though special welding techniques may be necessary in some cases (10).

1.10 Structural steel products

Hot steel ingots can be formed into a variety of structurally useful shapes by passing them through a succession of rolling mills which progressively reduce the original bulk material.

Figure 1.8 shows a range of the shapes (or *steel sections* as they are more commonly known).

Plate and strip steel produced by hot rolling can be formed into a variety of fabricated sections by shaping and welding. Amongst the products made in this way are hollow sections in the form of rectangles, squares and circles (Figure 1.9).

Thin plate or strip can be formed without heating into a wide range of cold-

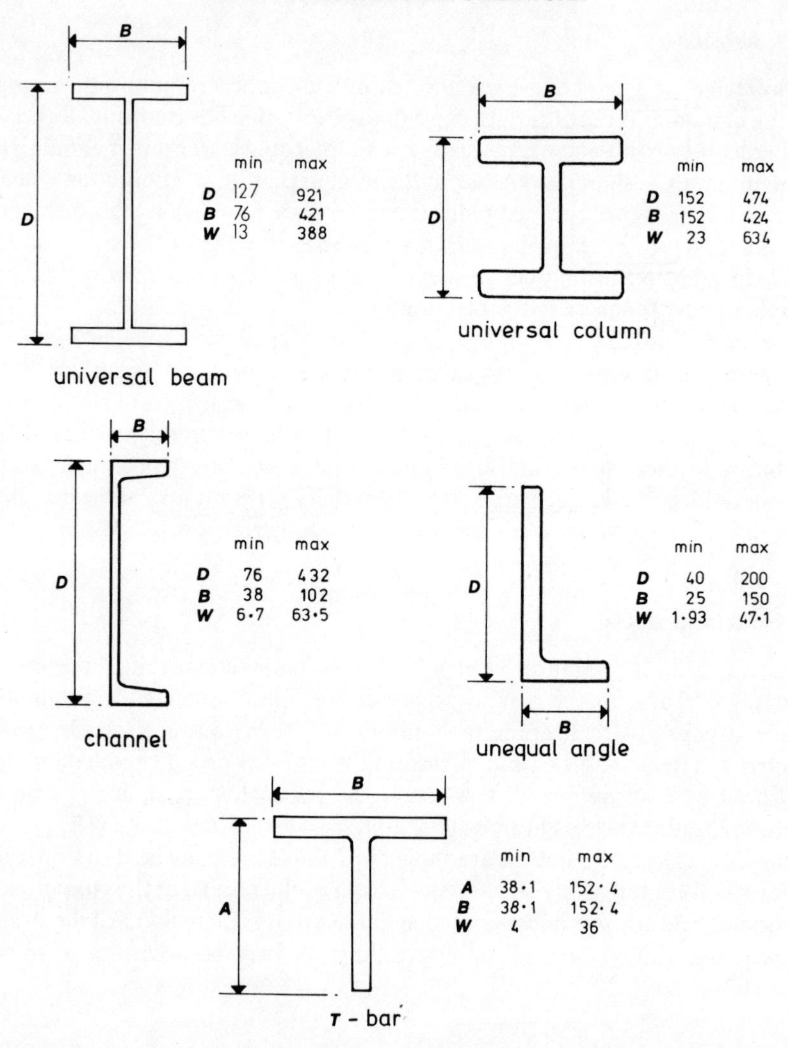

Figure 1.8 Hot-rolled steel sections (Dimensions in mm are the minimum and maximum available in the British Standard range. W = mass in kg of 1 m length)

rolled sections of considerable complexity (Figure 1.10). Cold-rolled sections have advantages in lighter forms of construction where the hot-rolled sections would be excessively strong (11).

Amongst other standard fabricating techniques is the method of increasing the depth of a rolled beam by *castellating* (12). The technique is illustrated in Figure 1.11(a). A zig-zag line is cut along the beam web by an automatic flame cutting machine. The two halves thus produced are rearranged so that the teeth match up and the teeth are then welded together. An even greater

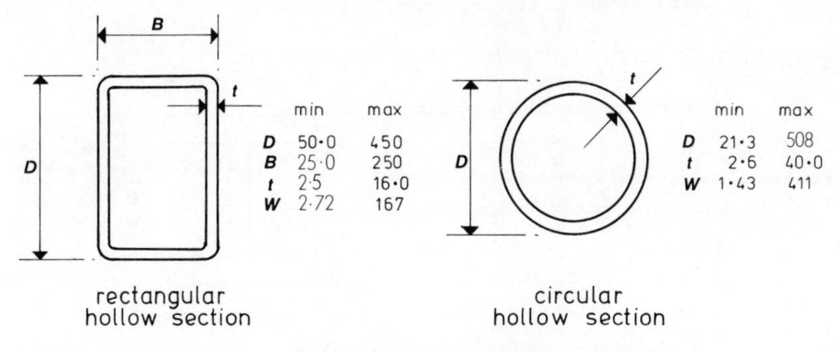

Figure 1.9 Hollow sections

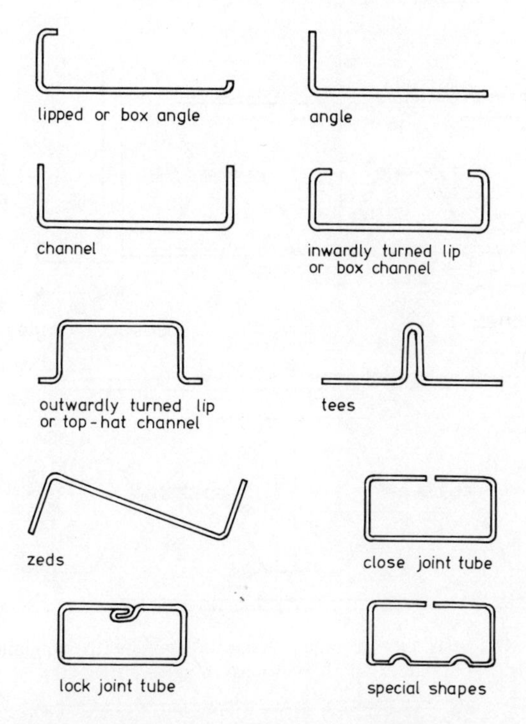

Figure 1.10 Cold-rolled steel sections. (In general, almost any shape can be produced by cold roll forming)

expansion is made possible by the insertion of a plate between the teeth (Figure 1.11(b)).

Automatic or semi-automatic fabricating methods are applied to the production of welded plate girders which consist of two plate flanges welded to a plate web. Girders with equal or unequal flanges can be welded without

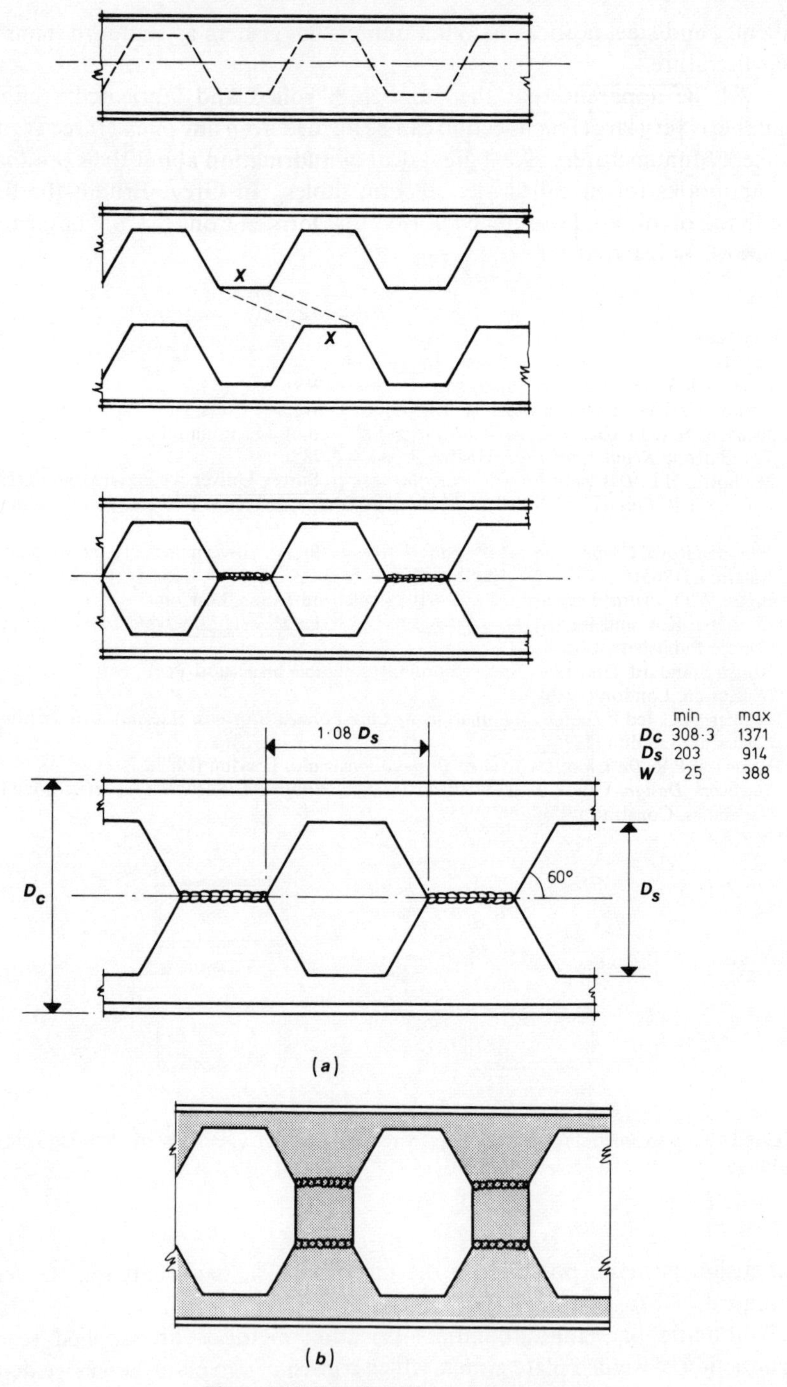

Figure 1.11 Castellated beam fabrication

difficulty and steel fabricators often quote such girders as standard items in their literature.

It will be apparent that the number of rolled and fabricated sections available is very large (each section can be formed from any one of three grades of steel). Manufacturers give a great deal of information about their products in handbooks (often called steel section tables). In Great Britain the first handbook of this kind was issued in 1887 by Dorman Long & Co. The current handbook is listed at reference (13).

References

1. Gale, W.K.V. *Iron and Steel.* Longman, London (1969).
2. Condit, Carl W. *American Building Art.* Oxford University Press, New York (1960).
3. Jackson, N. (ed.). *Civil Engineering Materials* (3rd edn.). Macmillan, London (1983).
4. *The Building Regulations 1985.* HMSO, London (1985).
5. Malhotra, H.L. *Design of Fire-Resisting Structures.* Surrey University Press, London (1982).
6. Gurney, T.R. *Fatigue of Welded Structures* (2nd edn.). Cambridge University Press, London (1979).
7. *Report of Royal Commission into the Failure of Kings Bridge.* Government Printer Melbourne, Australia (1963).
8. Biggs, W.D. *Brittle Fracture of Steel.* Macdonald and Evans, London (1960).
9. Chandler, K.A. and Bayliss, D.A. *Corrosion Protection of Steel Structures.* Elsevier-Applied Science Publishers, London (1985).
10. British Standard 4360:1986. Specification for Weldable Structural Steels. British Standards Institution, London (1986).
11. Walker, A.C. (ed.). *Design and Analysis of Cold-Formed Sections.* International Textbook Company, London (1975).
12. Knowles, P.R. *Design of Castellated Beams.* Constrado, London (1985).
13. *Steelwork Design.* Guide to BS5950: Part 1:1985. Volume 1. Section Properties. Member Capacities. Constrado, London 1985.

2 Design and stability

2.1 Design

Structural design is a process by which a structure required to perform a given function is proportioned to satisfy certain performance criteria (size, shape, etc.) in a safe and economic way. To a large extent design is a pencil and paper exercise (electronic computation must also be included in the list of design tools), by which a mathematical model of the real structure is tested for adequate performance. It is only for very large or novel structures that testing on physical models or full-size prototypes is carried out. Thus, the designer is very much more circumscribed than for example his counterpart in aeronautical or marine engineering, where prototype testing is commonplace.

The great difficulty in teaching design is that the subject is entirely open-ended. To any given structural performance specification there is an infinity of solutions which will at least satisfy the safety criterion although many will clearly be uneconomic. But this latter aspect of economy—the cheapest structure to perform the given function—has no easy answer. It is almost impossible to predict the most economic solution; except for very simple schemes the best that can be done is to select some promising alternative solutions which can then be priced. In many cases the most straightforward will be the cheapest because contractors may be wary of any unusual or novel structural system; in this way a price restraint is put on innovation. One point which must be made is that solutions aimed at using the smallest amount of material (minimum weight designs) are very often not the most economical.

2.1.1 *Safety*

There are in practice two aspects of a structure which are not easily quantified: the strength of the materials from which it is constructed and the magnitude of the loads which it must resist. A designer's duty is to proportion the structure in such a way that the risk of its failing is acceptably small. In order to carry out this task he must have adequate information about the probability of the materials having a strength below some given datum (*characteristic strength*) or of loads exceeding a given intensity (*characteristic load*). Of the two it is clear that he has more control over the question of material strength; estimation of imposed load intensity is much less exact (though self weight can be calculated accurately). To the beginner the most difficult aspect is the estimation of the self weight (dead load) of the structure for, as he rightly observes, until the structure is designed the dead load is unknown but until the dead load is known the structure cannot be designed. A circular argument of this kind can be resolved by making a guess at the dead load (based on, for example, a similar scheme), and then checking the dead load of the resulting structure. If the initial guess is not grossly different then a second calculation should be sufficient. The problem should not, in any case, be over emphasized; for modest size structures the dead load is only a small portion of the total loading (1).

The imposed (live) loading arises from a number of sources, not only external to the structure (stored material, snow, people, wind) but also from internal effects such as temperature differential. Standard loading intensities are given in Codes of Practice (2). There will generally be a number of possible combinations of loading types so that it may be necessary to investigate more than one case to determine the critical combination.

The properties of structural steel (particularly its yield stress) can be guaranteed by the steel maker in terms of a minimum value below which no test result will fall. The same cannot be said of concrete, a much more variable material, and so the calculation of the design strengths of these two materials must take this difference into account.

2.1.2 *Limit state design*

In recent years Codes of Practice have been written in *limit state* terms, a limit state being a condition of the structure which is unacceptable for one reason or another. Limit states may be classified into two main groups: catastrophic, involving for example total collapse of the structure, known as *ultimate limit states*, and less severe occurrences, such as excessive deflection or local yielding, known as *serviceability limit states*.

Examples of limit states which need consideration in the design of steel structures are:

Ultimate		*Serviceability*
Strength	Stability	Deflection
General yielding	Overturning	Vibration
Rupture	Sway	Fatigue damage which can be repaired
Buckling	Fatigue fracture	
Transformation into a mechanism	Brittle fracture	Corrosion and durability

In order to provide an acceptable probability against the attainment of a limit state appropriate factors (γ) must be applied to cover variations in:

Table 2.1 Partial safety factors for loads

Load system	Factor γ_f
Dead	
Generally	1.4
Restraining uplift or overturning	1.0
Acting with wind and imposed loads combined	1.2
Imposed	
Generally	1.6
Acting with imposed or crane load	1.2
Temperature effect forces	1.2

(a) material strength γ_m
(b) loading γ_l
(c) structural performance γ_p

In steelwork design γ_m is incorporated in the specified design strength of the material; γ_l and γ_p are not used independently but in terms of a factor γ_f the product of γ_l and γ_p. Values of γ_f are given in Table 2.1. Where loads occur in combination the probability of each load reaching a maximum value simultaneously is reduced.

The effect of loading on steel structures may be investigated with reference to two criteria of steel performance, elastic or plastic.

The basic premise of the *elastic design method* is that the attainment of the yield stress at any point in a structure marks the end of acceptable behaviour, the argument being that any further increase in stress will lead to permanent strain in the material. The designer has, therefore, to calculate the stresses in the structure, determine the maximum values and ensure that they are acceptable. Elastic design in limit state terms, therefore, sees the serviceability limit state of local yielding as the point at which structural adequacy ceases.

The criticism of elastic design made over 50 years ago was that measurement of the stresses in a real structure under working load revealed values which did not correspond with those calculated from an idealized mathematical model, because that model did not contain residual stresses, unforeseen joint stiffnesses or fabrication errors present in the real structure (3). A *plastic design* method was evolved which, taking account of the plastic extension of steel, was able to predict the load which would cause the structure to collapse (4).

2.2 Stability

2.2.1 *Instability of a compression number*

The satisfactory performance of a structure depends not only on its ability to withstand the loads imposed on it (resistance to rupture) but also on its remaining stable under these loads. Instability can take a number of forms:

(a) total instability of the structural system
(b) instability of a component in the system
(c) instability of an element forming part of a component.

Element and component stability problems may arise at points where there is partial or complete compressive load on the cross-section, conditions which must occur somewhere in all practical structures.

An instructive approach to the problem is first to consider the behaviour of an ideal model and then to modify this behaviour to take account of the nature of real structures. A suitable starting point is the ideal strut, absolutely straight, of uniform section, of linear elastic material having infinitely large yield stress and pinned supports at each end (Figure 2.1), the subject of Leonhard Euler's analysis published in 1744. The action of this strut under increasing load is illustrated in Figure 2.2. It can be shown that the *critical buckling load*

$$F_{cr} = \frac{\pi^2 EI}{L^2}$$

At this load the strut will buckle, the load will become eccentric and so the cross section will be stressed not only in compression but also in bending. This additional stress is of no consequence if the strut material has infinite strength

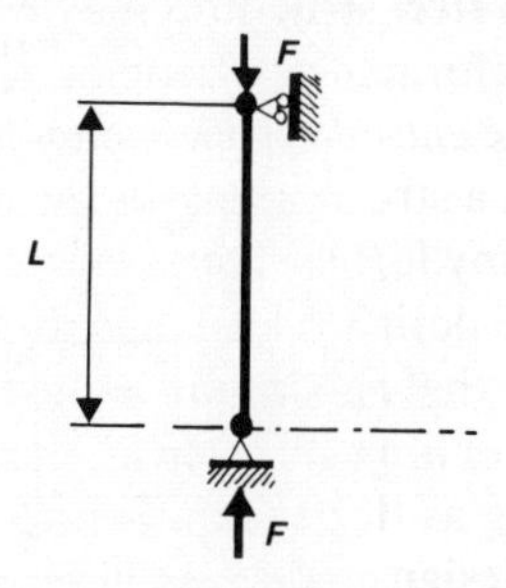

Figure 2.1 The Euler strut

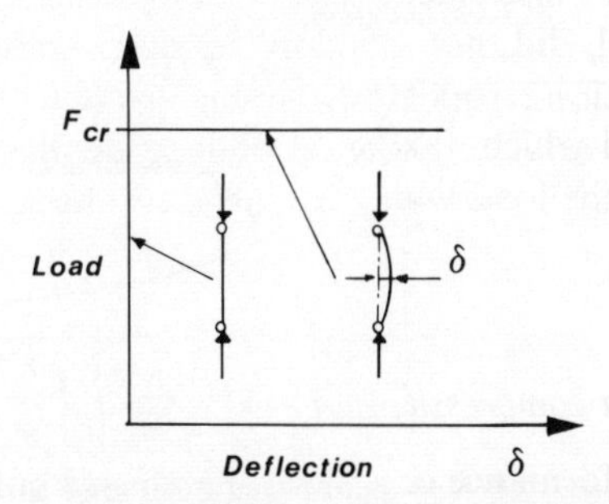

Figure 2.2 Euler strut: load and lateral deflection

but for an elastic–plastic material with a defined yield stress increasing lateral deflection will eventually lead to compressive yield in the strut.

Noting that

$$\lambda = \frac{L}{r}, \quad f = \frac{F}{A}, \quad r^2 = \frac{I}{A}$$

the critical buckling stress

$$f_{cr} = \frac{\pi^2 E}{\lambda^2}$$

Figure 2.3 shows how the critical stress is related to the slenderness ratio of an ideal Euler strut. Because the material has infinite strength this characteristic Euler hyperbola has infinite value at zero slenderness ratio.

For steels having an actual yield point and the idealised stress–strain curve of Figure 2.4a the Euler hyperbola is only valid for values of critical stress less than the yield stress. For greater values ($f_{cr} > f_y$) the strut will yield before buckling and so the curve will be modified to that of Figure 2.4b. A more

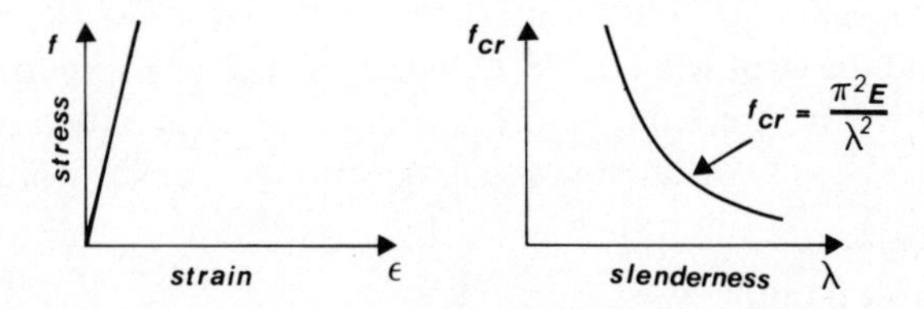

Figure 2.3 Euler strut of infinitely elastic material

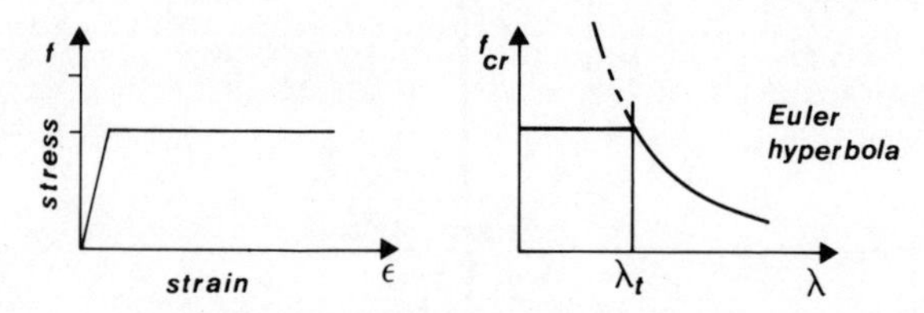

Figure 2.4 Euler strut of elastic–plastic material

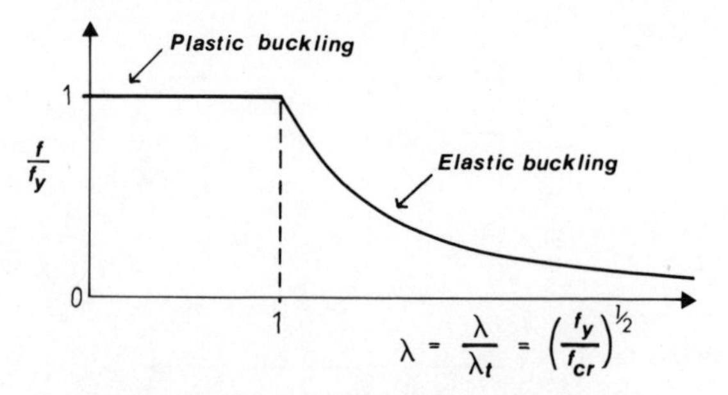

Figure 2.5 Non-dimensional Euler strut curve

useful form of this curve is produced by making the axes non-dimensional as shown in Figure 2.5; the vertical axis is the non-dimensional stress f/f_y and the horizontal axis is the non-dimensional slenderness

$$\bar{\lambda} = \frac{\lambda}{\lambda_t} = \left(\frac{f_y}{f_{cr}}\right)^{\frac{1}{2}}$$

where λ_t is the slenderness at which the yield plateau intersects the Euler hyperbola.

As the value of Young's modulus E is approximately the same for all grades of steel this single curve represents the behaviour of all struts having the idealised Euler and material properties described.

When the buckling stress is greater than the yield stress the maximum load that the strut can carry is termed the *squash load* $F_p = f_y A$. Conversely at greater values of slenderness the load will be restricted to $F = f_{cr} A$. At this load the strut will buckle, the load will become eccentric to the centroidal axis and the cross section will be stressed not only in compression but also in bending. The resistance to loading will then reduce as illustrated in Figure 2.6; the strut has no reserve of post buckling strength.

A real column differs from the Euler strut in three important respects:

(a) it is not axially straight because of inevitable manufacturing defects
(b) the load on it, even if specified as applied axially is, in practice, eccentric to the centroidal axis
(c) the material is not stress-free in the unloaded state because of the presence of residual stresses (see p. 108).

In addition regard must be given to the actual properties of the strut material.

Both *initial lack of straightness* and *eccentricity of loading* lead to the superimposition of a bending moment on the applied axial load.

Consider a strut having an initial lack of straightness $v_0 = a_0 \sin(\pi z/L)$ (Figure 2.7).

It can be shown (Appendix A) that the initial deflection will be increased

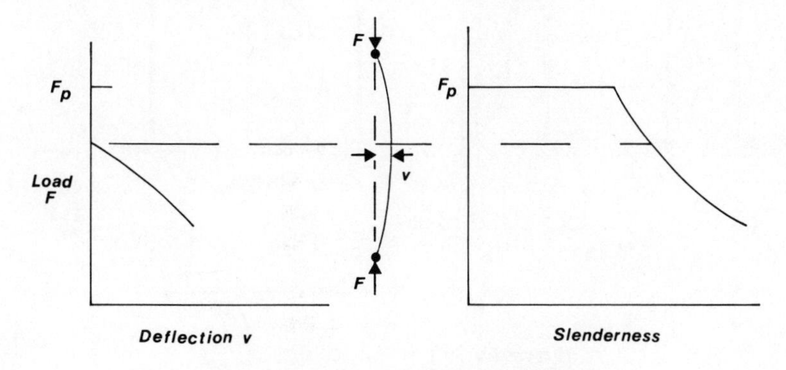

Figure 2.6 Post buckling behaviour of strut

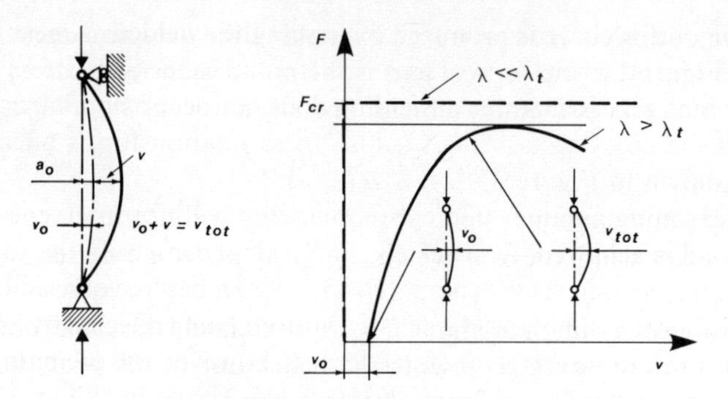

Figure 2.7 Strut with initial lack of straightness

under load F by the magnification factor

$$\frac{1}{\left(1-\frac{F}{F_{cr}}\right)}$$

In theory, therefore, as the applied load F approaches the buckling load F_{cr} the deflection tends to infinity. In fact the superimposition of a rapidly increasing bending moment on the axial compression will lead to compressive yield in the material at mid-height of the strut. The yielded zone spreads across the section and eventually the column collapses.

Residual stresses on a typical strut are shown in Figure 2.8a. The effect of a gradually increasing uniform stress is illustrated in Figures 2.8b and 2.8c. When $f_{av}+f_r=f_y$ (at a load F_{lim}) yielding starts at the flange tips and at the

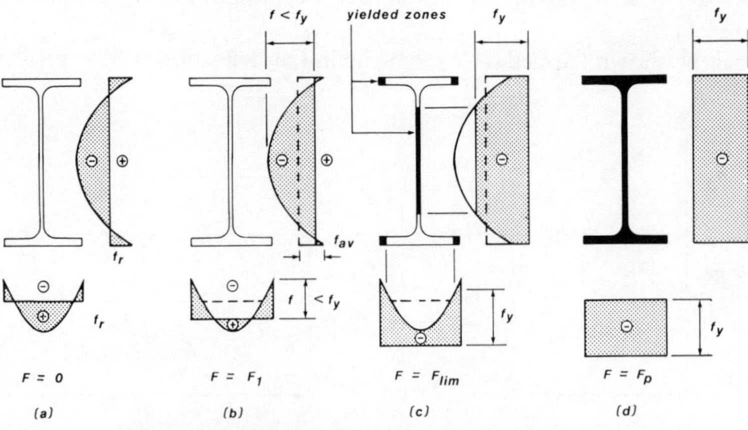

Figure 2.8 Strut with initial stresses
⊕ = tension

centre of the web. Increasing f_{av} causes the yielded zones to spread (Figure 2.8d) and when the average stress is equal to the yield stress the *squash load* has been reached. Because yielding does not occur simultaneously at all points in the cross section the load–stiffness relationship is affected in the manner shown in Figure 2.9.

This reasoning assumes that elastic buckling will not intervene before the squash load is achieved. A *stocky* ($F_{cr} \gg F_p$) strut can attain the squash load, albeit at an increased strain, compared with a member free of residual stress. A very slender strut will buckle elastically without being affected by any residual stress. Between these extremes of slenderness, however, the premature yielding produces loss of bending stiffness and leads to inelastic buckling at a load less than the elastic critical load. The effect is illustrated in Figure 2.10.

These abstract considerations indicate that the results of tests on real columns should fall below the theoretical Euler curve. Of the variable material properties only strain hardening will raise the predicted failure load above the squash load and then only at low slenderness ratios.

Stability, introduced here with reference to the simple strut, can be extended to plate elements and to the lateral torsional buckling of beams. The analytical manipulation is more complex and so is not considered in detail (there are

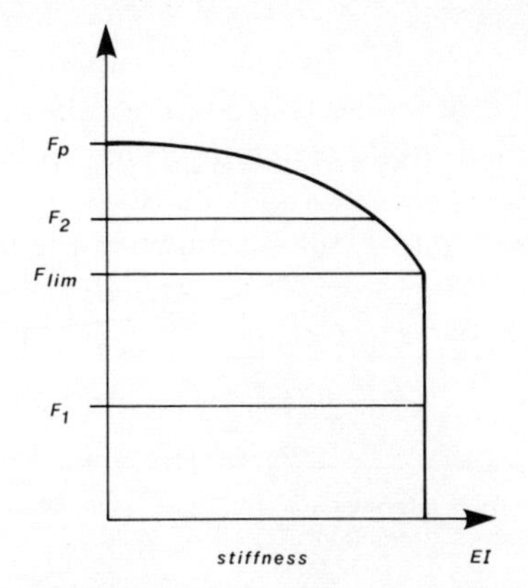

Figure 2.9 Load–stiffness relationship

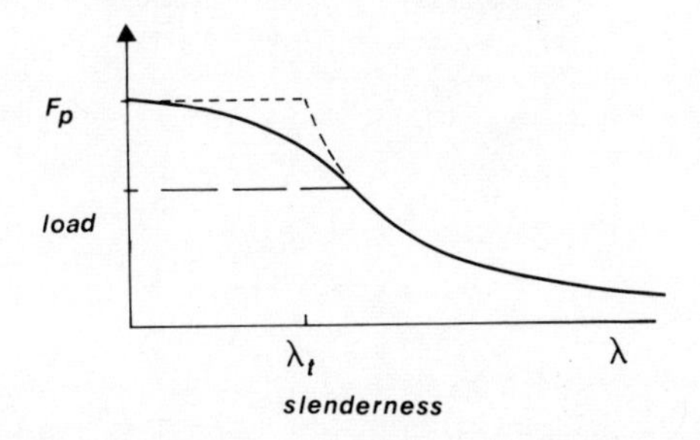

Figure 2.10 Strength curve for strut with residual stress

many good texts available (5) but the results form a basis for design of elements and beams.

In general all these elements will be affected by a physical characteristic (slenderness) which describes their resistance to instability. The extreme limits of behaviour are rupture on the one hand and elastic buckling on the other; the corresponding characteristics of the element are respectively *stocky* and *slender*. Between these two is an *intermediate transition zone* in which elastic and inelastic buckling interact. The shape of this general strength–slenderness curve is shown in Figure 2.11.

Plates The two-way action of plates means that elastic buckling stress is not the limit of their load carrying capacity; they exhibit post-buckling strength before collapse (Figure 2.12). Tests on real plates produce results similar to those for real columns but with collapse rather than elastic buckling as the limit of strength for slender plates.

Beams The stability of a beam is destroyed by the production of a lateral torsional buckle (Figure 2.13) at a critical bending moment M_{cr}. The limiting bending moment on a stocky beam is the plastic moment (analogous to the squash load of a strut).

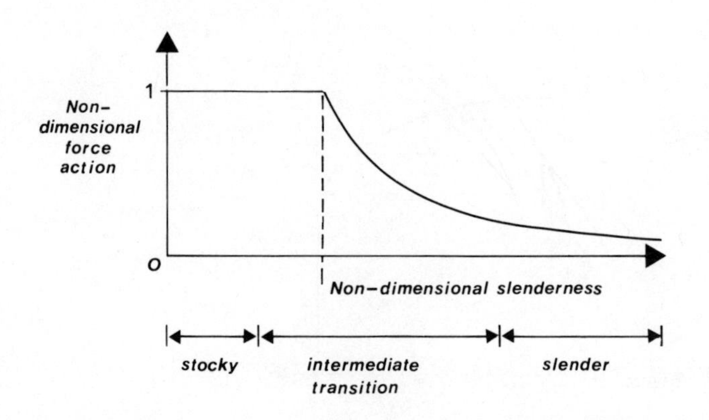

Member	Slenderness	Non-dimensional slenderness	Non-dimensional force action
Strut	$\frac{L_E}{r}$	$\left(\frac{f_y}{f_{cr}}\right)^{\frac{1}{2}}$	$\frac{F}{F_p}$
Beam	$\frac{L_E}{r}$	$\left(\frac{M_p}{M_{cr}}\right)^{\frac{1}{2}}$	$\frac{M}{M_p}$
Plate	$\frac{b}{t}$	$\left(\frac{f_y}{f_{cr}}\right)^{\frac{1}{2}}$	$\frac{F}{F_p}$

Figure 2.11 General ideal strength–slenderness curve

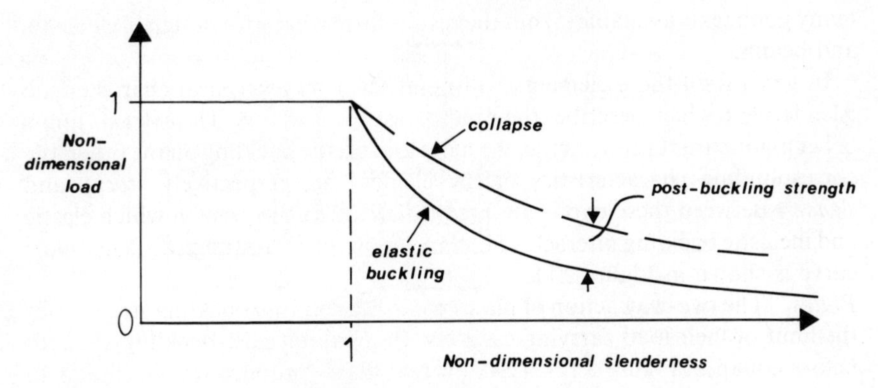

Figure 2.12 Strength of ideal flat plate

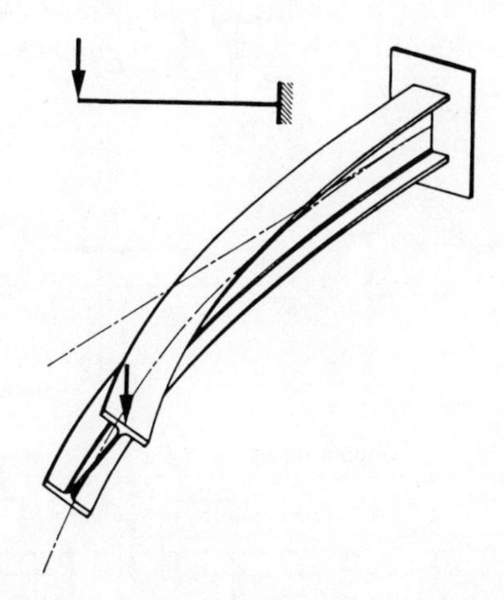

Figure 2.13 Lateral torsional buckling of a cantilever

The behaviour of beams and compression members is considered in detail in chapters 3 and 4 respectively.

2.2.2 *Local instability*

The elements (webs and flanges) composing a steel section are relatively slender; that is to say they have a thickness which is small in relation to their width and length. It is therefore possible for a web or flange to buckle prematurely at a load lower than that for the cross section considered as a whole causing a *local buckle* to form. Local buckling can be avoided by restricting the slenderness of beam elements.

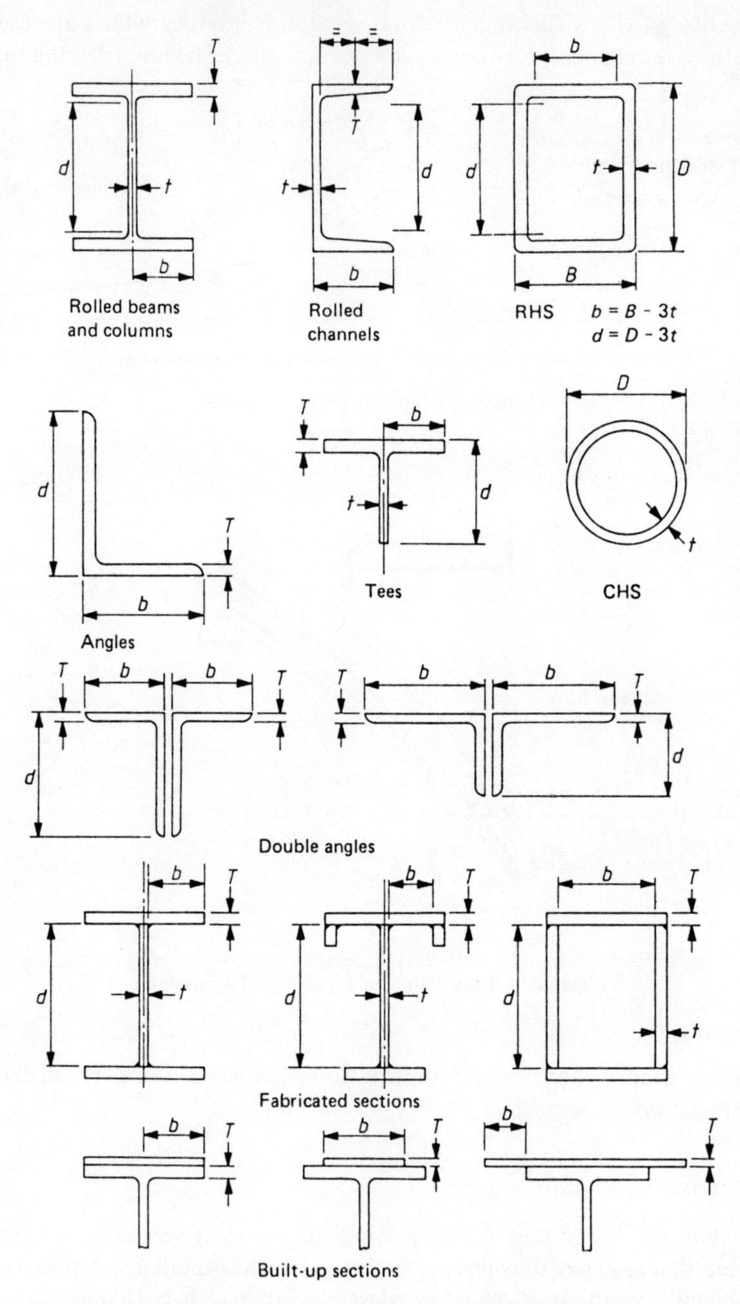

Figure 2.14 Dimensions for section classification (from BS 5950: Part 1: 1985)

Because of the wide range of cross-section geometries which are used in structures the elements forming a cross section are defined by referring to four classes:

(a) plastic
(b) compact
(c) semi-compact
(d) slender

Table 2.2 Classification of elements

Class 1 Plastic cross-sections. All elements subject to compression are plastic elements. Plastic cross-sections can develop plastic hinges which can rotate sufficiently to allow redistribution of moments in the structure. These sections are the only ones permitted in plastic design.
Class 2 Compact cross-sections. All elements subject to compression are compact elements. Compact cross-sections can develop their full plastic moment capacity but rotation of the plastic hinge may be restricted by local buckling thus making plastic design impossible.
Class 3 Semi-compact cross-sections. All elements subject to compression are semi-compact elements. Semi-compact sections can attain the design strength at the extreme fibres but the full plastic moment may not develop because of local buckling.
Class 4 Slender cross-sections. Elements subject to compression due to moment or axial load are slender elements. Slender cross-sections may not attain the design strength due to local buckling.

Table 2.3 Limiting width to thickness ratios

Element	Section	Ratio	Limiting ratio for $p_y = 275\ \text{N/mm}^2$ Class of element Plastic	Compact	Semi-compact
Compression flanges					
Outstand	Built-up	b/T	7.5	8.5	13.0
	Rolled	b/T	8.5	9.5	15.0
Internal	Built-up	b/T	23.0	25.0	28.0
	Rolled	b/T	26.0	32.0	39.0
Webs					
Neutral axis at mid-depth	All	d/t	79.0	98.0	120.0
Generally	All	d/t	$\frac{79.0}{0.4 + 6\alpha}$	$\frac{98.0}{6\alpha}$	
Compressed throughout	Built-up	d/t			28.0
	Rolled	d/t			39.0

1. Dimensions b, d, t, T are defined in Figure 2.14.
2. $\varepsilon = (275/p_y)^{1/2}$.
3. For steels having a yield stress p_y other than $275\ \text{N/mm}^2$ multiply tabulated limiting values by ratio ε.
4. $\alpha = 2y_c/d$, where y_c is the distance from the plastic neutral axis to the edge of the web connected to the compression flange. But if $\alpha > 2.0$ the section should be taken as having compression throughout.
5. Sections are 'built-up' by welding.

The limiting width-to-thickness ratios for the different classes are given in Table 2.2. Figure 2.14 defines the element dimensions. A cross section is classified as plastic, compact, semi-compact or slender by referring to the classes of element composing it. A cross-section may contain more than one class of element in which case it is the classification of the most slender which governs its own classification.

References

1. Blockley, D.I. *The Nature of Structural Design and Safety*. Ellis Horwood, Chichester (1980).
2. British Standard BS 6399: Design Loadings for Buildings, Parts 1, 2 and 3. British Standards Institution, London 1984.
3. Baker, J.F. *The Steel Skeleton*. Vol. 1, *Elastic Behaviour and Design*. Cambridge University Press, London (1954).
4. Baker, J.F., Horne, M.R. and Heyman, J. *The Steel Skeleton*. Vol. 2, *Plastic Behaviour and Design*. Cambridge University Press, London (1956).
5. Allen, H.G. and Bulson, P.S. *Background to Buckling*. McGraw-Hill Maidenhead (1980).

3 Beams

3.1 General

Beams have been used for many centuries but their systematic design had to await the development of a theory of bending. Such intellectual giants as Leonardo da Vinci and Galileo concerned themselves with the strength of beams but it was not until nearly 200 years after Galileo's death that Navier derived the correct flexural stress formula. Torsional stresses and lateral buckling were investigated by late nineteenth and twentieth century workers.

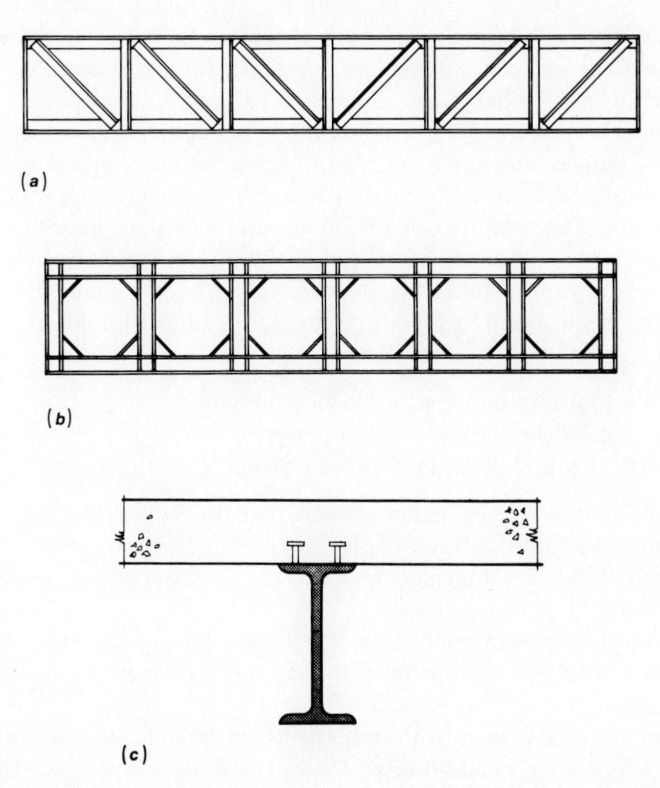

Figure 3.1 Types of beam (*a*) Lattice girder; (*b*) Vierendeel truss; (*c*) composite beam

Ultimate load theories which interested Leonardo da Vinci and Galileo were given coherence only in recent years; indeed work is still proceeding in this field.

Structural elements which primarily resist bending are known as *beams* or *girders*, the former term being commonly applied to rolled sections and the latter to fabricated members. Very small beams are often called *joists*. In addition to solid web and castellated beams there are open web flexural members of the triangulated lattice type shown in Figure 3.1a and of the rigid jointed Vierendeel type shown in Figure 3.1b. The loads in the components of a lattice girder are predominantly compressive or tensile and so their design is considered elsewhere (the stress distribution in Vierendeel girders, being complex, is beyond the scope of this publication). Where concrete decks are supported by steel beams the two may be interconnected to form a composite beam (Figure 3.1c).

3.2 Efficiency

The efficient utilisation of material in a beam is determined by the geometrical layout of web and flanges. If bending is the only criterion then the best solution is a beam in which all the material is concentrated in the flanges and the flanges are separated as far apart as possible. In practice there will be need for some web material to keep the flanges apart and to resist shear. As a measure of beam efficiency it is possible to relate the allocation of a given amount of material to flanges and web to satisfy three different and generally mutually contradictory criteria of elastic bending strength, plastic bending strength and beam stiffness. The formulae which follow can be used as guides to relative proportions where the beam designer has freedom to vary the web and flange sizes; subject always to restrictions concerned with local instability.

Consider the idealized beam in Figure 3.2. The dimension h is assumed to represent any one of the actual dimensions:

(a) height d of web between inside faces of flanges
(b) distance h between centroids of flanges
(c) overall depth D of beam between outside faces of flanges.

The justification for this assumption is that the flange thickness is small when compared with the beam depth.

The web depth to thickness ratio $k = h/t$ is determined generally by considerations of web stability.

Component cross-sectional areas: A_w = web area, A_f = total area of *both* flanges, A = total area of beam $= A_f + A_w$, and α = web area/total beam area $= A_w/A$.

It is now possible to express the section properties in terms of k, α and A by use of the following substitutions: $A_w = \alpha A$ and $A_w = ht = h^2/k$. Therefore $h = (k\alpha A)^{\frac{1}{2}}$ and $A_f = (A - A_w) = A(1 - \alpha)$.

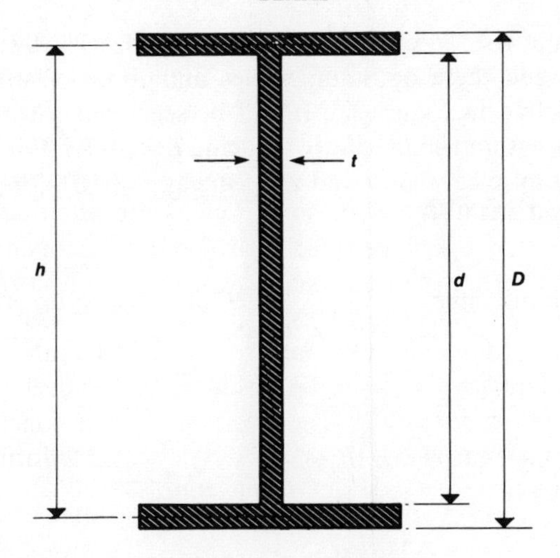

Figure 3.2 Idealized beam

Second moment of area I

$$I = A_f\left(\frac{h}{2}\right)^2 + \frac{th^3}{12}$$

$$= \frac{A(1-\alpha)k\alpha A}{4} + \frac{\alpha A k\alpha A}{12}$$

$$= \frac{k\alpha A^2(3-2\alpha)}{12}$$

Elastic section modulus Z

$$Z = I/(h/2)$$

$$= \frac{k\alpha A^2(3-2\alpha)}{12} \times \frac{2}{(k\alpha A)^{\frac{1}{2}}}$$

$$= \frac{(k\alpha)^{\frac{1}{2}}A^{\frac{3}{2}}(3-2\alpha)}{6}$$

Plastic section modulus S

$$S = 2\left(\frac{A_f}{2} \times \frac{h}{2}\right) + \frac{th^2}{4}$$

$$= \frac{A(1-\alpha)(k\alpha A)^{\frac{1}{2}}}{2} + \frac{\alpha A(k\alpha A)^{\frac{1}{2}}}{4}$$

$$= \frac{(k\alpha)^{\frac{1}{2}}A^{\frac{3}{2}}(2-\alpha)}{4}$$

These expressions may be differentiated with respect to α to determine the relative area of flange and web to give the optimum value of I, Z or S, and the corresponding depth h for some given steel area A, as shown below. The relationship between these optimum values and those occurring at other values of α is plotted in Figure 3.3. It will be seen that α can vary within well spaced limits without significantly reducing I, Z and S. For elastic design a suitable compromise between I and Z occurs at $\alpha = 0.63$, for plastic design between I and S at $\alpha = 0.60$.

Optimum value	Optimum occurs at	
	α	h
$I = \frac{3}{32}kA^2 = 0.0937\,kA^2$	$\frac{3}{4}$	$0.87\,(kA)^{\frac{1}{2}}$
$Z = \frac{1}{3\sqrt{2}}k^{\frac{1}{2}}A^{\frac{3}{2}} = 0.236\,k^{\frac{1}{2}}A^{\frac{3}{2}}$	$\frac{1}{2}$	$0.71\,(kA)^{\frac{1}{2}}$
$S = \frac{2^{\frac{1}{2}}}{3^{\frac{3}{2}}}k^{\frac{1}{2}}A^{\frac{3}{2}} = 0.272\,k^{\frac{1}{2}}A^{\frac{3}{2}}$	$\frac{2}{3}$	$0.82(kA)^{\frac{1}{2}}$

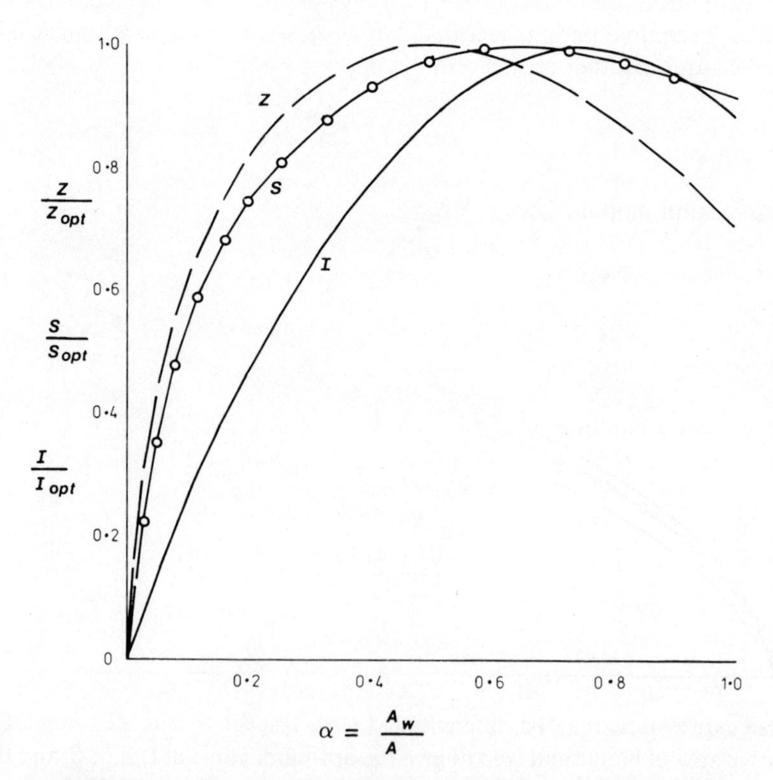

Figure 3.3 Variation of section properties with change in relative web area

By rearrangement of the preceding formulae the optimum depth and steel area for a given I, Z or S can be written as:

Required Value	Optimum values A	Optimum values h
I	$\left(\frac{32}{3}\frac{I}{k}\right)^{\frac{1}{2}} = 3.27\left(\frac{I}{k}\right)^{\frac{1}{2}}$	$(6kI)^{\frac{1}{4}} = 1.56(kI)^{\frac{1}{4}}$
Z	$\left(18\frac{Z^2}{k}\right)^{\frac{1}{3}} = 2.62\left(\frac{Z^2}{k}\right)^{\frac{1}{3}}$	$\left(\frac{3}{2}kZ\right)^{\frac{1}{3}} = 1.14(kZ)^{\frac{1}{3}}$
S	$\left(\frac{27}{2}\frac{S^2}{k}\right)^{\frac{1}{3}} = 2.38\left(\frac{S^2}{k}\right)^{\frac{1}{3}}$	$(2kS)^{\frac{1}{3}} = 1.26(kS)^{\frac{1}{3}}$

The influence of the web depth to thickness ratio k on the optimum values is illustrated in Figure 3.4. Clearly the more slender the web the more efficient the beam but this advantage must be viewed in the light of the increased web stiffening required and the fact that very deep beams may be impractical for architectural or other reasons.

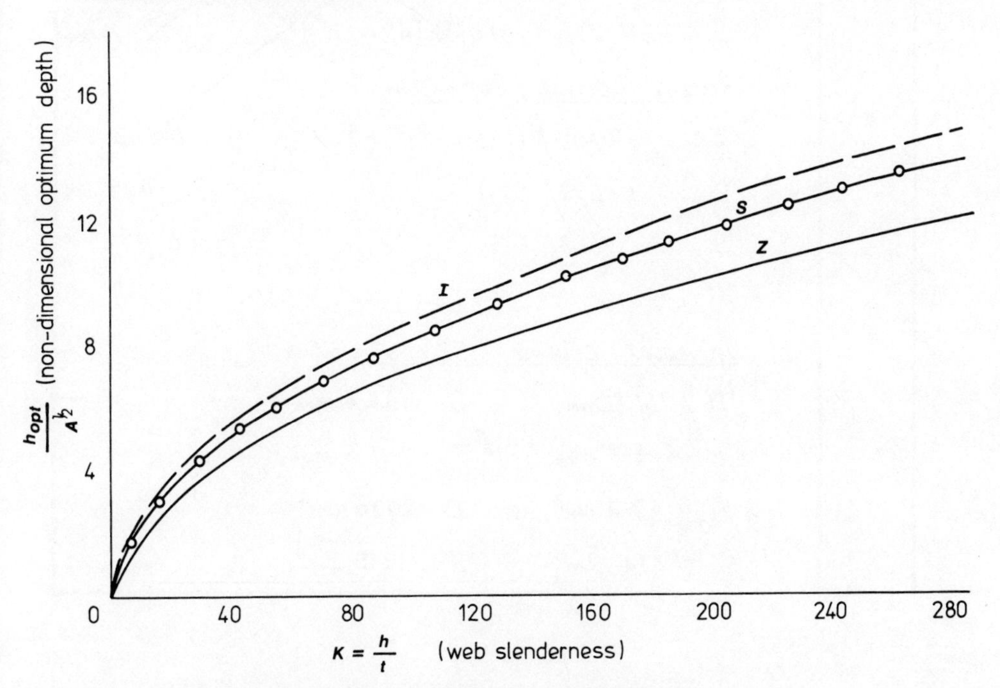

Figure 3.4 Influence of web slenderness on optimum section properties

BS5950 clause	Example 3.1 Efficiency of rolled sections as beams 3.1.1
	a. Universal beam 356 × 127 × 39
	D 352.8 mm B 126 mm
	t 6.5 mm T 10.7 mm
	A 49.4 cm² I 10100 cm⁴
	Z 572 cm³ S 654 cm³
	d 331.4 mm
	$A_w = 331.4 \times 6.5 = 2154\ mm^2$
	$\alpha = 2154/(49.4 \times 10^2) = 0.436$
	$k = 331.4/6.5 = 51.0$
	Optimum values:
	$I = 0.0937 \times 51.0 \times (49.4 \times 10^2)^2 = 117 \times 10^6\ mm^4$
	$Z = 0.235 \times (51.0)^{1/2} \times (49.4 \times 10^2)^{3/2} = 579 \times 10^3\ mm^3$
	$S = 0.272 \times (51.0)^{1/2} \times (49.4 \times 10^2)^{3/2} = 672 \times 10^3\ mm^3$
	Ratios actual/optimum:
	I 10100/11700 = 0.863
	Z 572/579 = 0.988
	S 654/672 = 0.973
	b. Universal column 152 × 152 × 37
	D 161.8 mm B 154.4 mm
	t 8.1 mm T 11.5 mm
	A 47.4 cm² I 2220 cm⁴
	Z 274 cm³ S 310 cm³

BS5950 clause	Example 3.2 Efficiency of rolled sections as beams 3.1.2
	d 138·8 mm
	$A_w = 138.8 \times 8.1 = 1124\ mm^2$
	$\alpha = 1124/47.4 \times 10^2 = 0.237$
	$k = 138.8/8.1 = 17.1$
	<u>Optimum values</u>:
	$I = 0.0937 \times 17.1 \times (47.4 \times 10^2)^2 = 36 \times 10^6\ mm^4$
	$Z = 0.235 \times (17.1)^{1/2} \times (47.4 \times 10^2)^{3/2} = 317 \times 10^3\ mm^3$
	$S = 0.272 \times (17.1)^{1/2} \times (47.4 \times 10^2)^{3/2} = 367 \times 10^3\ mm^3$
	<u>Ratios actual/optimum</u>:
	I 2220/3600 = 0·617
	Z 274/317 = 0·864
	S 310/367 = 0·845

The properties of rolled sections are of necessity a compromise but, as can be seen from Example 3.1, do not depart very much from the optimum values. The range of choice of rolled beams is large and by the nature of their mass production they tend to be relatively cheap and widely available. Thus some inefficiency in their cross-sectional layout can easily be tolerated. Welded plate girder profiles are under the designer's control; he can, within reason, adjust the depth and flange to web thickness ratio to give as economical a solution as possible.

3.3 Stability

The relatively slender profile of steel beams gives rise to a marked tendency to local or overall instability under certain conditions.

3.3.1 *Lateral instability*

Where a beam is provided with inadequate lateral restraint or has inadequate lateral stiffness it will fail (always assuming that its elements do not buckle prematurely), at a bending moment M_b lower than the full plastic moment by buckling sideways and twisting, a mode of failure known as *lateral torsional buckling* (Figure 3.5).

The slenderness of a beam, λ, is expressed as the ratio

$$\frac{\text{effective length}}{\text{minimum radius of gyration}} = \frac{L_E}{r_y}$$

in which L_E, the effective length, is based on the distance between lateral supports modified for the type of restraint given by these supports.

Where the slenderness is high the beam buckles at low stress in the elastic

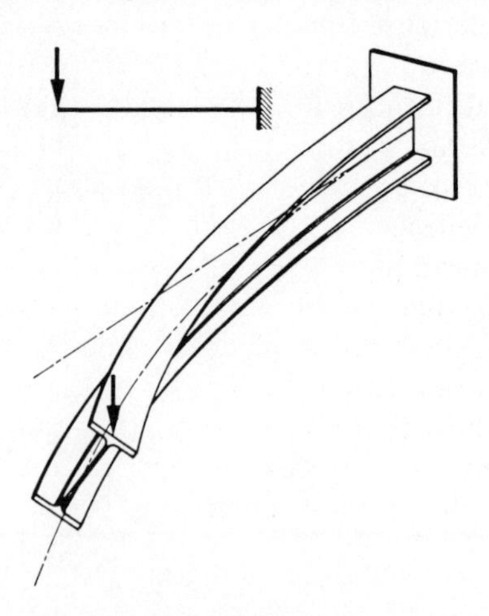

Figure 3.5 Lateral torsional buckling of a cantilever

range. It can be shown that the corresponding elastic buckling moment for a beam loaded by equal and opposite end moments is:

$$M_E = \frac{\pi}{L}\left(\frac{EI_y GJ}{\gamma}\right)^{\frac{1}{2}}\left(1 + \frac{\pi^2 EH}{L^2 GJ}\right)^{\frac{1}{2}}$$

where L = span, EI_y = minor axis flexural rigidity, GJ = torsional rigidity, EH = warping rigidity and $\gamma = (I_x - I_y)/I_x$ is an approximate allowance for the deflection in the plane of the web.

For low values of slenderness the beam does not fail in lateral torsional buckling but by the formation of a plastic hinge at the plastic moment M_p: such beams are described as 'stocky'. Between the extremes of low and high slenderness is a transition zone in which the mode of failure is by interactive lateral torsional buckling. The value of this failure moment is influenced by interaction between plasticity and instability, by residual stresses in the steel and by initial lack of straightness in the beam.

A numerical quantity which is useful in interpreting lateral instability data is the *equivalent slenderness*, λ_{LT}, in the expression

$$\frac{M_E}{M_p} = \frac{\pi^2 E}{p_y} \cdot \frac{1}{\lambda_{LT}^2}$$

Rearranging

$$\lambda_{LT} = \left(\frac{M_p}{M_E} \cdot \frac{\pi^2 E}{p_y}\right)^{\frac{1}{2}}$$

The advantage of using λ_{LT} for determining lateral buckling strength is that all the relevant properties of the material (E and G) and cross-section (I_y, H, J, and L) are included in this single parameter.

Because of the influence of residual stress and lack of straightness in beams used in construction, design recommendations must necessarily be related to the results of extensive test work. Experiments have been made on a wide range of beam types, fabricated from steels of different yield stress under varied loading conditions. A plot of some test results is shown in Figure 3.6. The horizontal axis of the plot uses a non-dimensional slenderness $\bar{\lambda} = (M_P/M_E)^{\frac{1}{2}}$ thus eliminating the effects of variations in yield strength between test specimens. Taking a mean line through the results shows that *stocky* beams can be defined as having $(M_P/M_E)^{\frac{1}{2}} < 0.4$ and *slender* beams as having $(M_P/M_E)^{\frac{1}{2}} > 1.2$. From these test results design curves, fitting the data, have been devised. They are based on the Perry formula, originally proposed in 1886 for columns and have the form $(M_E - M_b)(M_P - M_b) = \eta_{LT} M_E M_b$, where η_{LT} is a Perry coefficient which represents imperfections in the beam. Its value can be adjusted to fit the plot in a suitable fashion:

(a) rolled beams

$$\eta_{LT} = 0.007(\lambda_{LT} - \lambda_{LO})$$

nor less than zero

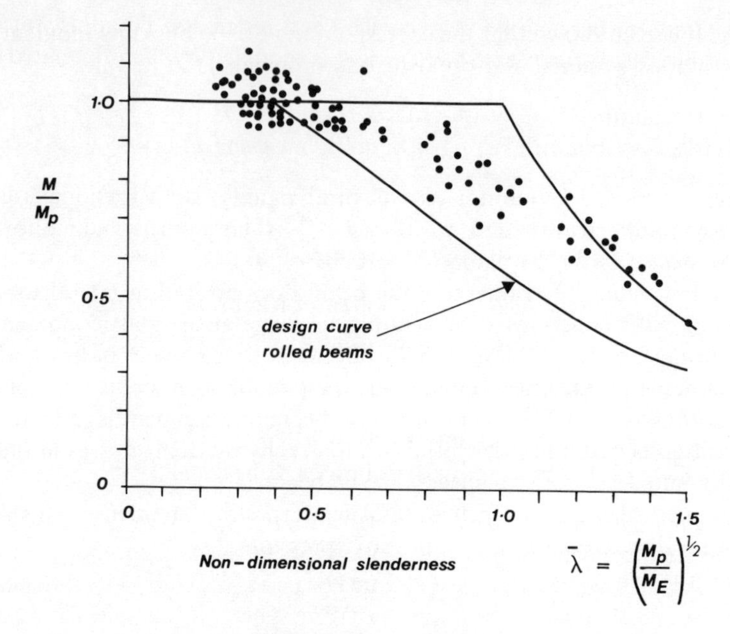

Figure 3.6 Tests on beams

where

λ_{LO}, the limiting equivalent slenderness to develop M_p, is

$$0.4(\pi^2 E/p_y)^{\frac{1}{2}}.$$

(b) welded beams, which because of different, and more deleterious, patterns of residual stress show lower buckling strengths than rolled sections:

$$\eta_{LT} = 0.014\lambda_o$$

but

$$\leqslant 0.014\,(\lambda_{LT} - \lambda_{LO})$$

$$\geqslant 0.007\,(\lambda_{LT} - \lambda_{LO})$$

nor less than zero.

Solving the Perry formula for M_b

$$M_b = M_E M_p / [\phi_B + (\phi_B^2 - M_E M_p)]^{\frac{1}{2}}$$

where

$$\phi_B = [M_p + (\eta_{LT} + 1)M_E]/2$$

For design purposes it is more convenient to work in terms of a *buckling resistance stress* p_b such that M_b = plastic section modulus times p_b. Values of p_b for rolled and welded sections can be obtained from Tables 3.11 and 3.12 respectively of BS5950. These values refer to uniform symmetrical beams under uniform bending having specified end restraints. To adjust for other conditions, the actual beam slenderness is modified by factors related to

(a) the nature of the beam cross-section (u, v)
(b) the distribution of applied bending moment (m or n).
(c) the nature of lateral restraint.

Nature of cross section
A *slenderness factor* v is based on two sectional properties

$$N = \frac{\text{second moment of area of compression flange}}{\text{second moment of area of tension flange}}$$

both calculated about the minor axis and

$$\frac{\lambda}{x} = \frac{\text{minor axis slenderness of beam}}{\text{torsional index of beam}}$$

A *buckling parameter* u, is solely a section property. Both these parameters can be calculated or, more readily, obtained from tables. Conservatively u may be taken as 0.9 for rolled I, H or channel beams and 1.0 for any other beams; v may exceed 1.0 in some cases of unsymmetrical beams with smaller compression than tension flanges but otherwise may be taken conservatively as 1.0. The application of these factors is illustrated in Example 3.2.

Distribution of applied bending moment
The case of constant bending moment assumed in deriving p_b may be related to the actual bending moment distribution on a beam in one of two ways:

(a) by adjusting the actual bending moment by an equivalent uniform moment factor m
(b) by adjusting the slenderness of the beam by a slenderness correction factor n.

Conservatively both m and n may be taken as 1.0. Otherwise Table 3.1 gives, for members of uniform cross-section, the conditions under which lower values from BS5950 of m or n (but not both) may be used.

Table 3.1 Use of correction factors m and n for members not subject to destabilizing load*

Member condition	m	n
Loaded in length between adjacent lateral restraints	1.0	Table 15 or 16
Not loaded in length between adjacent lateral restraints	Table 18	1.0
Cantilevers	1.0	1.0

* *Note*: For all members subject to destabilizing loads $m = n = 1.0$.

BS5950 clause	Example 3.2 Properties of a rolled beam 3.2.1
	457 x 152 x 52 UB
	Tabulated values:
	D 449.8 mm B 152.4 mm t 7.6 mm
	T 10.9 mm A 66.5 cm^2 r 10.2 mm
	I_x 21300 cm^4 I_y 645 cm^4 Z_x 949 cm^3
	Z_y 84.6 cm^3 S_x 1090 cm^3 S_y 133 cm^3
	r_{yy} 3.11 cm
	Torsion constant J
B2.5.1(c)	$J = \frac{1}{3}(t_1^3 b_1 + t_2^3 b_2 + t_w^3 h_w)$
	$t_1 = t_2 = 10.9$, $b_1 = b_2 = 152.4$
	$t_w = 7.6$, $h_w = 449.8 - 2 \times 10.9 = 428.0$
	$J = \frac{1}{3}(2 \times 10.9^3 \times 152.4 + 7.6^3 \times 428)$
	$= 194.2 \times 10^3$ mm^4 $= 19.4$ cm^4
	[tabulated value, 21.3 cm^4, takes account of flange to web radii]
	Warping constant H
B2.5.1(c)	$H = \dfrac{h_s^2 t_1 t_2 b_1^3 b_2^3}{12(t_1 b_1^3 + t_2 b_2^3)}$
	$h_s = 449.8 - 10.9 = 438.9$ mm
	$H = \dfrac{438.9^2 \times 10.9^2 \times 152.4^6}{12(2 \times 10.9 \times 152.4^3)} = 310 \times 10^9$ mm^6
	$= 0.310$ dm^6

BS5950 clause	Example 3.2 Properties of a rolled beam 3.2.2
	Buckling parameter u
B2.5.1(b)	$u = \left[\dfrac{I_y S_x^2 \gamma}{A^2 H}\right]^{1/4}$
	$\gamma = 1 - \dfrac{I_y}{I_x} = 1 - \dfrac{645}{21300} = 0.9697$
	$u = \left[\dfrac{645 \times 10^4 \times 1090^2 \times 10^6 \times 0.9697}{66.5^2 \times 10^4 \times 310 \times 10^9}\right]^{1/4} = 0.858$
	Torsional index x
B2.5.1(b)	$x = 1.132 \left[\dfrac{A H}{I_y J}\right]^{1/2}$
	$= 1.132 \left[\dfrac{66.5 \times 310 \times 10^{11}}{645 \times 194.2 \times 10^7}\right]^{1/2} = 45.9$
	Approximations
4.3.7.5 4.3.7.7	$u = 0.9$, $x = \dfrac{D}{T} = \dfrac{449.8}{10.9} = 41.3$

BS5950 clause	Example 3.2 Properties of a rolled beam	3.2.3
	Slenderness factor v	
B2.5.1(d)	$$v = \left[\left(4N(1-N) + \frac{1}{20}\left(\frac{\lambda}{x}\right)^2 + \psi^2 \right)^{1/2} + \psi \right]^{-1/2}$$	
	$$N = \frac{I_{cf}}{I_{cf} + I_{tf}} = 0.5$$	
	Calculate v for $\lambda = 190$:	
	$\psi = 0$ for $N = 0.5$	
	$$v = \left[\left(4 \times 0.5(1-0.5) + \frac{1}{20} \times \frac{190^2}{45.9^2} \right)^{1/2} \right]^{-1/2}$$	
	$= 0.857$	
	[table 14 contains values of v for a range of values of N and $\frac{\lambda}{x}$]	

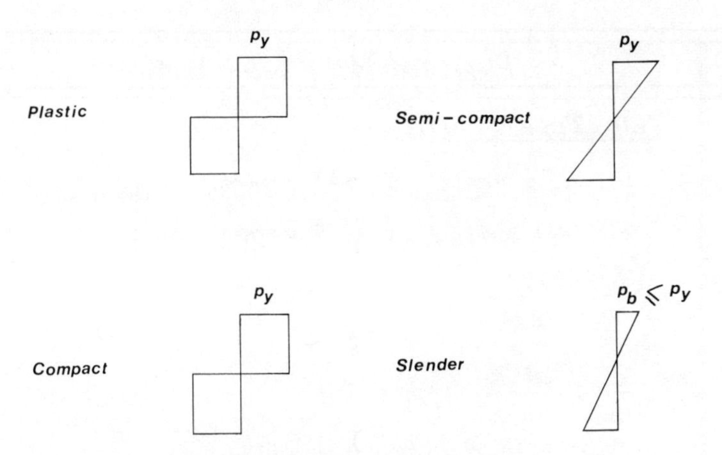

Figure 3.7 Permitted stress distributions

3.3.2 *Local instability—flanges*

The division of plate elements into four classes related to their outstand-to-thickness ratios is explained in Section 2.2.2. The moment capacity of a beam may be limited by the onset of premature flange buckling if its flange falls into class 3 or 4; only class 1 sections are suitable for plastic design.

The permitted stress distributions for the four classes are illustrated in Figure 3.7. It is only for slender elements that complication arises. In this case the section design strength p_y is reduced by the stress reduction factor given in Table 3.2 and the reduced design strength is used throughout the design of the section.

3.3.3 *Local instability—webs*

Although a slender beam cross-section with a symmetrical profile has a web d/t greater than 120ε there is a further restriction that when $d/t > 63\varepsilon$ the web should be checked for shear buckling. This restriction does not affect rolled

Table 3.2 Stress reduction factors for slender sections

Element	Section type	Stress reduction factor*
Outstand of compression flange	Built-up	$10/[(b/T\varepsilon) - 3]$
	Rolled	$11/[(b/T\varepsilon) - 4]$
Internal of compression flange	Built-up	$21/[(b/T\varepsilon) - 7]$
	Rolled	$31/[(b/T\varepsilon) - 8]$

* *Note:* $\varepsilon = (275/p_y)^{\frac{1}{2}}$

BS5950 clause	Example 3.3 Classification of cross section	3.3.1
3.5	610 × 229 × 101 UB	
	B 227·6 mm d 547·3 mm	
	T 14·8 mm t 10·6 mm	
Figure 3	$\frac{b}{T} = \frac{227 \cdot 6}{2 \times 14 \cdot 8} = 7 \cdot 69$	
	$\frac{d}{t} = \frac{547 \cdot 3}{10 \cdot 6} = 51 \cdot 6$	
Table 6	material thickness ≤ 16 mm	
Table 7 note 3		

Grade	p_y N/mm²	ε
43	275	1·0
50	355	0·88
55	450	0·78

Limiting width to thickness ratios:

Table 7 — a. outstand element of compression flange b/T

Grade	Element class 1	2	3
43	8·5	9·5	15·0
50	7·5	8·4	13·2
55	6·6	7·4	11·7

BS5950 clause	Example 3.3 Classification of cross section
3.3.2	

b. web, neutral axis at mid depth d/t

Grade	Element class 1	2	3
43	79.0	98.0	120.0
50	69.5	86.2	105.6
55	61.6	76.4	93.6

Element classifications:

Grade	web	flange
43	plastic	plastic
50	plastic	compact
55	plastic	semi-compact

Cross section classification:

43	plastic
50	compact
55	semi-compact

Table 3.3 Elastic critical shear stress

a/d	Critical shear stress q_e
$\leqslant 1.0$	$\left(0.75 + \dfrac{1.0}{(a/d)^2}\right)\left(\dfrac{1000}{d/t}\right)^2$
> 1.0	$\left(1.0 + \dfrac{0.75}{(a/d)^2}\right)\left(\dfrac{1000}{d/t}\right)^2$

Table 3.4. Critical shear strength of a web panel.

Web slenderness λ_ω	Type	Critical shear strength q_{cr}
$\leqslant 0.8$	Stocky	$0.6p_y$
$\geqslant 1.25$	Slender	q_e
$0.8 < \lambda_w < 1.25$	Transition	$0.6p_y[1 - 0.8(\lambda_w - 0.8)]$

$\lambda_w = (0.6p_y/q_c)^{\frac{1}{2}}$, p_{yw} = design strength of the web

universal beams or columns which all have webs less slender but will usually be operative for plate girders fabricated from relatively thin plates.

It is at points of high shear force that critical web shear buckles will occur. The shear buckling resistance of a panel is $V_{cr} = dtq_{cr}$, the value to be given to q_{cr} being decided by using the now familiar idea of stocky, transitional or slender panel action based on shear yield, interactive elastic–plastic buckling or elastic buckling respectively. The elastic critical stress q_e for a panel loaded in shear is dependent on two parameters

(a) the ratio of the panel sides a/d, known as the aspect ratio,
(b) the width to thickness ratio d/t.

Table 3.3 gives the relationships between these parameters and the elastic critical shear stress. q_{cr} can be calculated from Table 3.4 or read directly from Tables 21a to 21d in BS5950. $\lambda_w = (0.6p_y/q_e)^{\frac{1}{2}}$ and p_{yw} is the design strength of the web.

The aspect ratio of a normal unstiffened beam web is taken as infinite for the purpose of calculating q_{cr}. If such a web has insufficient shear resistance it may be strengthened by division into a series of panels of finite aspect ratio by suitable *intermediate vertical stiffeners* (Figure 3.8). Leaving aside the question of the proportions of stiffeners required to divide the web into a series of panels it is apparent from Table 3.3 that aspect ratios lying between 0.5 and 2.0 will allow the most efficient use of material. Selection of the actual stiffener spacing is then decided on a practical basis so as to divide the web into suitable panels.

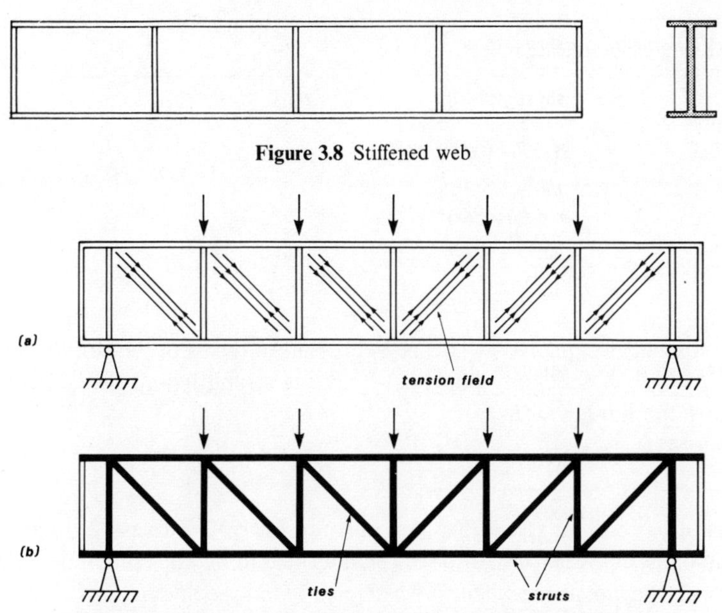

Figure 3.8 Stiffened web

Figure 3.9(*a*) Tension field in stiffened web; (*b*) analogous *N* truss

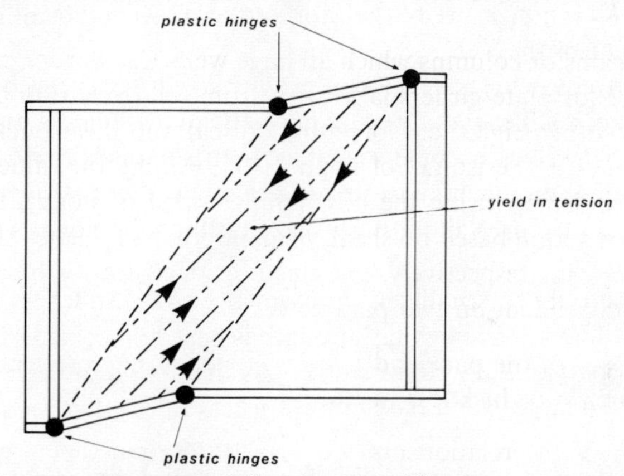

Figure 3.10 Failure in web and flange

The actual stiffness is determined from empirical rules based on tests which ensure that panel buckling is forced between the stiffeners.

Tests on stiffened plate girders have shown that the simple approach outlined above is conservative in that it takes no account of the *post buckling strength* of the web panels, known as *tension field action.* This phenomenon is illustrated in Figure 3.9; the web panels between stiffeners act as diagonal tension members in a manner analogous to an *N* truss. Failure eventually occurs when the web yields in tension and plastic hinges form in the flanges (Figure 3.10).

Table 3.5 Minimum stiffener stiffness

Actual spacing of stiffener a	Minimum stiffness I_s
$< 1.41d$	$1.5d^3t^3/a^2$
$\geqslant 1.41d$	$0.75dt^3$

The full shear capacity V_b can be expressed in terms of the sum of the web shear strength q_b (buckling plus post-buckling strength) and the contribution made by the flanges $q_f(K_f)^{\frac{1}{2}}$

$$V_b = q_b + q_f(K_f)^{\frac{1}{2}}$$

Values of basic shear strength q_b are given in BS5950. Tables 22a, b and c for a range of aspect ratio (a/d) and web slenderness (d/t). The contribution of the flanges, q_f, the *flange dependent shear strength factor*, comes from Tables 23a, b and c for the same ranges of aspect ratio and web slenderness. Finally the parameter K_f is a measure of the ratio of the plastic moment capacity of the flange to that of the web.

Vertical (transverse) stiffeners are provided to force the web to buckle into panels between stiffeners and so require adequate *stiffness*. In addition they must not themselves become unstable. If these stiffeners do not carry any external load or moment the necessary stiffness will be provided if the stiffener second moment of area about the web centreline, I_s, is not less than the values in Table 3.5.

Where there are external loads or moments on the stiffeners they must have increased stiffness. The buckling strength is based on considering the stiffener acting with part of the web, as a strut. As the design of the various types of stiffener is to a large extent based on empirical rules reference should be made to Example 3.10.

Another form of web instability can occur under a concentrated load or reaction, the web acting as a compression member and the load being assumed to disperse uniformly into the web at 45° through half the depth of the section. Referring to Figure 3.11, the average compressive stress at middepth is

$$\frac{P}{b_1 + 2n_1}$$

If this stress does not exceed the strength of a strut of slenderness ratio $2.5d/t$ then a load bearing stiffener is not required. Otherwise load bearing stiffeners welded to either side of the web are employed. They form, with a portion of the

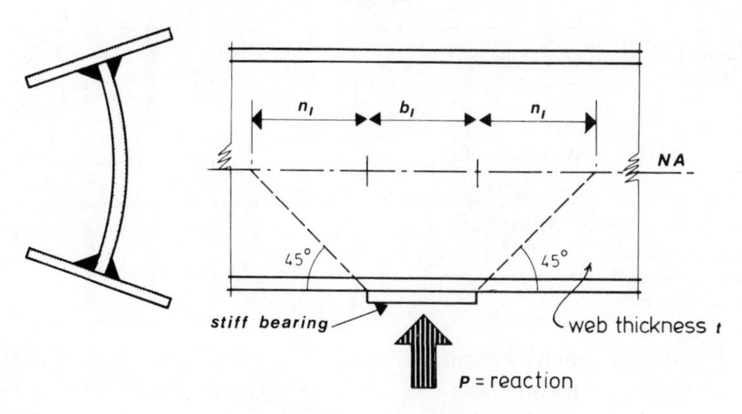

Figure 3.11 Web buckling

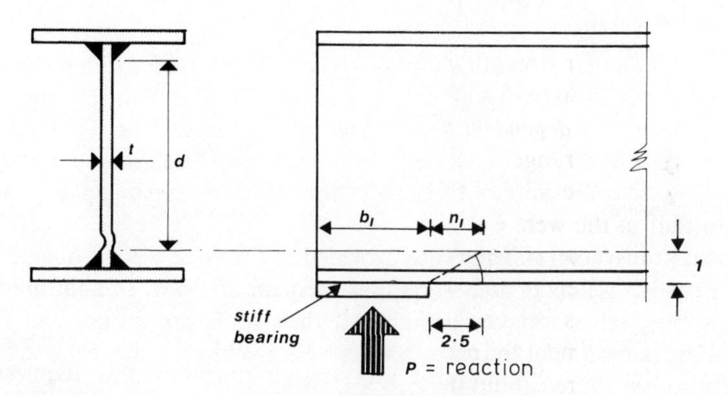

Figure 3.12 Web crushing

web, a cross-shaped strut having a much lower slenderness than the web alone (see Example 3.11).

Although not a form of stability it is convenient to mention here the phenomenon of web crushing which can occur under concentrated loads. The critical point of high stress occurs at the end of the root radius or fillet weld illustrated in Figure 3.12, dispersion of load being assumed to be at 1 to 2.5 to the horizontal from the ends of the bearing plate. Where the load exceeds the local capacity of the web bearing stiffeners must be provided (see Example 3.12).

3.4 Shear

The interaction between shear and bending moment is such that the moment capacity is only significantly affected when the shear force is greater than 0.6 times the shear capacity.

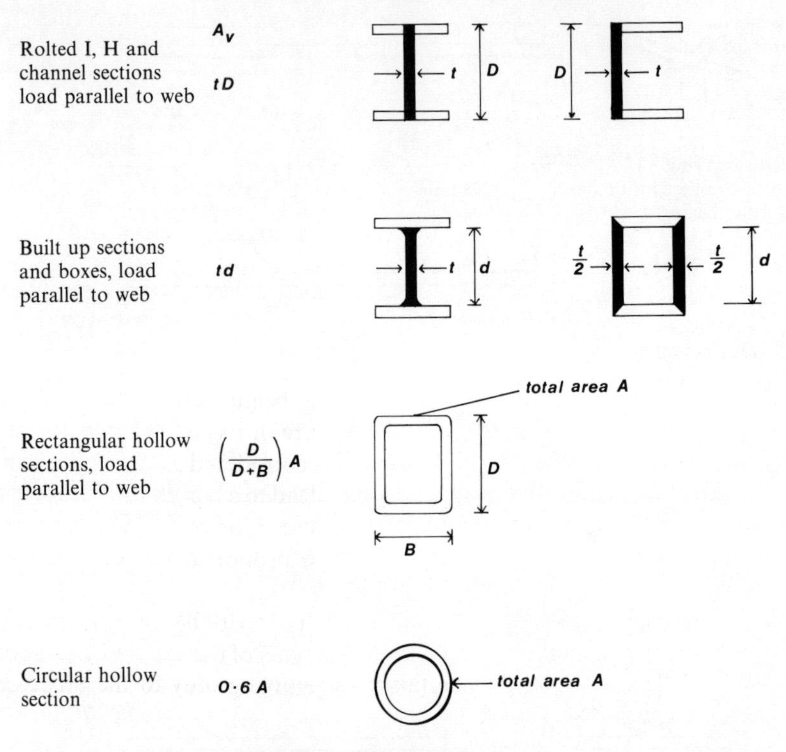

Figure 3.13 Shear areas

The shear capacity of a section is calculated from $P_v = 0.6p_yA_v$, where A_v is the shear area (see Figure 3.13). For design purposes this requirement commonly means that the interaction between shear force and bending moment is not important and so the full moment capacity may be used. If the shear force is greater than $0.6P_v$:

(a) for plastic or compact sections the plastic modulus S is reduced by a factor which takes account of the web bending capacity which is committed to resisting the applied shear force. The reduced plastic modulus is $(S - S_v\rho_1)$, where $\rho_1 = (2.5F_v/P_v) - 1.5$ and S_v is:

  (i) for sections with equal flanges:
  the plastic modulus of the shear area i.e. rolled sections $tD^2/4$, built up sections $td^2/4$.

  (ii) for sections with unequal flanges:
  the plastic modulus of the gross section minus the plastic modulus of that part of the section remaining after deduction of the shear area.

(b) For semi-compact and slender sections $M_c = p_yZ$, where p_y is the design strength reduced if required for slender sections and Z is the relevant elastic section modulus.

Table 3.6 Deflection limits

Beam type	Deflection limit
Cantilever	Length/180
Beams carrying plaster or other brittle finish	Span/360
All other beams	Span/200

3.5 Deflection

An important serviceability criterion is that beam deflections should not impair the strength or efficiency of the structure or its components nor cause damage to finishes. Deflection should also be considered excessive if it can be perceived by the general public as it may then lead to a lack of confidence in the structure even if there is no risk of collapse. Commonly the maximum permitted vertical deflection is expressed as a proportion of the span of the beam (Table 3.6).

Permanent dead load deflection will not change with time; the beam can be initially cambered upward to counteract all or part of the dead load deflection. Thus the deflection limits can be taken as applying only to the unfactored imposed loading.

3.6 Approach to beam design

The process is essentially one of trial and error. A beam is selected having a plastic moment of resistance somewhat greater than that actually required to resist the maximum applied bending moment M. The buckling resistance moment M_b is then determined. If $M_b \geqslant M$ then the initial selection is adequate; otherwise further trials are made. Once a satisfactory section has been obtained checks can be made on the adequacy of the web in buckling, shear and bearing and of the beam in deflection.

Determination of the buckling resistance moment has to take account of the following factors:

(a) the nature of the beam

- (i) uniform I sections with equal flanges
- (ii) uniform I sections with unequal flanges
- (iii) box sections
- (iv) non-uniform I sections
- (v) cantilevers

(b) the distribution of applied bending moment
(c) the conditions of lateral support

Table 3.7 Moment capacity at low shear load for sections of each class with either full or other lateral restraint

Moment capacity at low shear load[1]		
	Lateral restraint	
Section class	Full	Other cases
Plastic[2]	$p_yS \leqslant 1.2p_yZ$	$p_bS \leqslant 1.2p_yZ$
Compact	$p_yS \leqslant 1.2p_yZ$	$p_bS \leqslant 1.2p_yZ$
Semi-compact	p_yZ	$p_bS \leqslant p_yZ$
Slender	p_yZ (Note 3)	$p_bS \leqslant p_yZ$

Notes
1. Low shear load is when the shear force does not exceed 0.6 times the shear capacity of the section.
2. These are the only sections for which a plastic analysis of the structure is permitted
3. p_y is reduced by a stress reduction factor as necessary to take account of the slenderness of the cross-section—see Table 3.2.

The major factors affecting the method of design are:

(a) the section classification
(b) the conditions of lateral restraint

Section classification will determine whether

(a) plastic or elastic distribution of stress is to be assumed in the cross section
(b) in the case of indeterminate beams a plastic analysis may be used to determine the distribution of bending moment

In all cases, unless the beam has full lateral restraint, the lateral torsional bukling resistance of the member must be checked.

It must be emphasized that the value of p_b is related to a beam which is:

(a) simply supported
(b) of constant cross-section throughout its length
(c) loaded by a constant bending moment
(d) laterally restrained only at its supports and then only against rotation about its longitudinal axis (i.e. torsion) and lateral deflection.

Conditions of restraint, bending moment distribution and section variation which differ from these will affect the bending strength p_b. See Examples 3.4 and 3.5.

BS5950 clause	Example 3.4 Buckling resistance moment of UB 3.4.1
	<u>457 x 152 x 52 UB</u>
	Steel grade 43
	Section properties page 3.2.1
Table 6	p_y = 275 N/mm² (T < 16 mm
Table 7	Plastic section : b/T = 6·99 < 8·5
	d/t = 53·6 < 79·0
	Moment capacity (low shear load)
4.2.5	M_c = 275 x 1090 x 10^{-3} = **300 kN m**
	(1·2 $p_y Z$ = 1·2 x 275 x 949 x 10^{-3} = 313 kN m)
	Assume effective length 5·0 m
4.3.7.1	a. <u>Conservative approach</u>
4.3.7.7	λ = 5 x 10^2/3·11 = 161
	x = D/T = 41·3
Table 19(b)	p_b = 88·6 N/mm²
	M_b = 88·6 x 1090 x 10^{-3} = **96·6 kN m**
4.3.7.3	b. <u>More accurate approach</u>
	From tables u = 0·859 , x = 43·9
	λ/x = 161/43·9 = 3·67
	N = 0·5
Table 14	v = 0·88
4.3.7.5	λ_{LT} = 0·859 x 0·88 x 161 = 122 (n = 1·0)
Table 11	p_b = 93·6 N/mm²
	M_b = 93·6 x 1090 x 10^{-3} = **102 kN m**

BS5950 clause	Example 3.5 Use of factors m and n 3.5.1
	457 × 152 × 52 UB
	steel grade 43 5.0m
Table 9	Effective length L_E assumed 5.0m
	Lateral restraints at supports
	a. Use of factor n
	Loading - uniformly distributed load between adjacent lateral supports
Table 13	$m = 1.0$, n from Table 15 or 16
	M=0 M_o βM=0
	$\gamma = M/M_o = 0$
	$\beta = 0$
Table 16	$n = 0.94$
	$\lambda_{LT} = (5000/31.1) \times 0.94 \times 0.859 \times 0.88 = 114$
	[u and v from example 3.4]
Table 11	$p_b = 103\ N/mm^2$
	$M_b = 103 \times 1.09 = \boxed{112\ kNm}$
	b. Use of factor m
	lateral restraint at load points
	265kN 25kN A B C D 1.5m 2.0m 1.5m
	145 kNm 290

BS5950 clause	Example 3.5 Use of factors m and n 3.5.2
	Effective lengths AB = CD = 1.5m
	BC = 2.0 m
4.2.5	M_{cx} = 300kNm > 290kNm
	check buckling resistance of AB, BC, CD

	L_E (m)	λ	λ/x	v	λ_{LT}	p_b (N/mm²)	M_b (kNm)
AB	1.5	48.2	1.10	0.986	40.8	260	283
BC	2.0	64.3	1.47	0.971	53.6	229	250

Table 13 — n = 1.0, m from Table 18

A-B

β = 0

Table 18 — m = 0.57

$\bar{M}$ = 290 × 0.57 = 165kNm < 283kNm

βM = 0

M = 290

B-C

β = 0.5

Table 18 — m = 0.76

$\bar{M}$ = 290 × 0.76 = 220kNm < 250kNm

βM = 145

M = 290

C-D is, by inspection, adequate

3.6.1 *Restrained compact beams*

Some simplification can be made by restricting consideration initially to beams composed of elements which permit the beam cross section to be classified as compact and with lateral restraint effective enough to permit the full plastic moment of the section to be utilized. Such a beam is described as 'restrained compact'. The design of restrained compact beams will require selection of a cross-section adequate in:

(a) bending: with a plastic moment of resistance not less that the applied bending moment (Table 3.7)
(b) shear
(c) amplitude of deflection
(d) web strength in bearing and buckling.

To provide this degree of lateral restraint the equivalent lateral slenderness λ_{LT} must not exceed the *limiting equivalent slenderness* λ_{LO} given by $\lambda_{LO} = 0.4 (\pi^2 E/p_y)^{\frac{1}{2}}$ tabulated in Table 3.8.

Although the accurate value of λ_{LT} is given by

$$\lambda_{LT} = \frac{M_p}{M_E}\left(\frac{\pi^2 E}{p_y}\right)^{\frac{1}{2}}$$

a useful approximation can be made by multiplying the minor axis slenderness by the product *nuv*, where *u* and *v* are properties of the section and *n* is a slenderness correction factor which depends on the distribution of applied bending moment. The calculations involved in determining λ_{LT} may be

Table 3.8 Limiting equivalent slenderness

p_y, Nmm^2	245	265	275	325	340	355	415	430	450
λ_{LO}	36	35	34	32	31	30	28	27	27

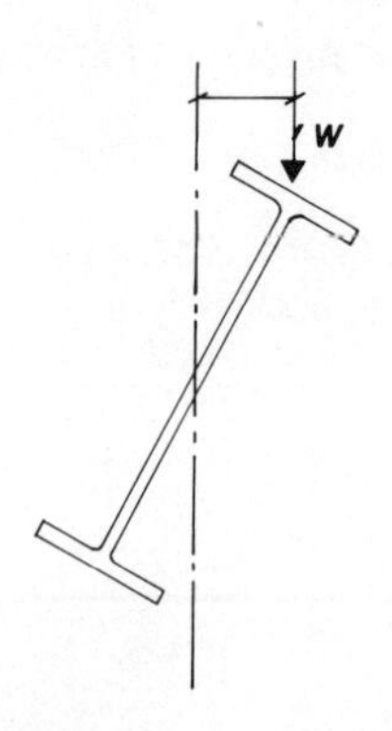

Figure 3.14 Destabilising load

BS5950 clause	Example 3.6 Bending and web capacities 3.6.1
	610 × 305 × 238 UB
	B 311·5 mm T 31·4 mm
	d 537·2 mm t 18·6 mm
	Grade 50 steel
3.1.1	Design strength
Table 6	$16 < T < 63$, $p_y = 340\ N/mm^2$
3.5	Section classification
	$\varepsilon = (275/340)^{1/2} = 0{\cdot}899$
	$b = 311{\cdot}5/2 = 155{\cdot}8$ mm
	$b/T = 155{\cdot}8/31{\cdot}4 = 4{\cdot}96$
	$d/t = 537{\cdot}2/18{\cdot}6 = 28{\cdot}9$
Table 7	limits for plastic section :
	flange $8{\cdot}5 \times 0{\cdot}899 = 7{\cdot}64$ $(> 4{\cdot}96)$
	web $79 \times 0{\cdot}899 = 71{\cdot}0$ $(> 28{\cdot}9$
	Section is plastic
Table 7, note 2.	$d/t < 63\varepsilon$ – no check on shear buckling needed.
	Moment capacity (low shear load)
	$S_x = 7460\ cm^3$
4.2.5	$M_{cx} = 7460 \times 340 \times 10^{-3}$ = 2540 kNm
B2.4	Limiting equivalent slenderness λ_{LO}
	$n = 1{\cdot}0$, $u = 0{\cdot}886$, $r_y = 7{\cdot}22$ cm , $x = 21{\cdot}1$

BS5950 clause	Example 3.6 Bending and web capacities 3.6.2
	$\lambda_{LO} = 0.4\left(\frac{\pi^2 E}{p_y}\right)^{1/2} = 0.4\left(\frac{\pi^2 \times 205 \times 10^3}{340}\right)^{1/2} = 30.9$
	$\lambda/x = 30.9/21.1 = 1.47$
Table 14	$v = 0.97$
	$L_E = \frac{30.9 \times 7.22 \times 10}{1.0 \times 0.886 \times 0.97} = 2600$ mm
	ie M_c is 2540 kNm provided that adequate lateral restraints are spaced at no more than 2.60 m
	<u>Lateral torsional buckling resistance moment for an effective length of 10.0m</u>
	$\lambda = 10.0 \times 10^3 / 7.22 \times 10 = 138.5$
	$\lambda/x = 138.5/21.1 = 6.56$
Table 14	$v = 0.75$
	$\lambda_{LT} = 138.5 \times 1.0 \times 0.886 \times 0.75 = 92.0$
Table 11	$p_b = 154$ N/mm^2
	$M_b = 154 \times 7460 \times 10^{-3} =$ **1150 kNm**
4.3.7.7	<u>Simplified method using Table 19</u>
	$\lambda = 138.5$, $x = 21.1$, $n = 1.0$
	Inter-polate for p_b:

BS5950 clause	Example 3.6 Bending and web capacities 3.6.3

λ \ x	20	21·1	25
135	158	154·7	143
138·5		151·2	
140	153	149·7	138

$M_b = 151{\cdot}2 \times 7460 \times 10^{-3} = \boxed{1130\ \text{kNm}}$

4.2.3 **Shear capacity**

Shear area $A_v = 18{\cdot}6 \times 633{\cdot}0 = 11{\cdot}8 \times 10^3\ \text{mm}^2$

Shear capacity $P_v = 11{\cdot}8 \times 10^3 \times 0{\cdot}6 \times 340 \times 10^{-3}$

$= \boxed{2400\ \text{kN}}$

Web buckling values

Beam factor C_1

For beam without flange plate

4.5.2.1 $n_1 = \frac{D}{2} = 633{\cdot}0/2 = 316{\cdot}5\ \text{mm}$

$\lambda = 2{\cdot}5\,(d/t) = 2{\cdot}5 \times 537{\cdot}2/18{\cdot}6 = 72{\cdot}2$

Table 27c $p_c = 205\ \text{N/mm}^2$

BS5950 clause	Example 3.6 Bending and web capacities 3.6.4
	area at neutral axis = $316.5 \times 18.6 = 5.89 \times 10^3 \text{mm}^2$
	$C_1 = 205 \times 5.89 \times 10^3 \times 10^{-3}$ = **1210 kN**
	Stiff bearing factor C_2
	For 1mm length of stiff bearing:
	area at neutral axis = $1 \times 18.6 = 18.6 \text{mm}^2$
	$C_2 = 18.6 \times 205 \times 10^{-3}$ = **3.81 kN/mm**
	Flange plate factor C_3
	For 1mm thickness of flange plate:
	area at neutral axis = $1 \times 18.6 = 18.6 \text{mm}^2$
	$C_3 = 18.6 \times 205 \times 10^{-3}$ = **3.81 kN/mm**
	Web bearing values
	T+r; b_1; n_2; 1; 2.5; flange plate; stiff bearing
	Beam factor C_1
	For beam without flange plate:
	$n_2 = 2.5(T+r) = 2.5(31.4+16.5) = 119.8 \text{mm}$
	area at end of radius = 119.8×18.6

BS5950 clause	Example 3.6 Bending and web capacities 3.6.5
	$= 2{\cdot}23 \times 10^3\ mm^2$
	$C_1 = 2{\cdot}23 \times 10^3 \times 340 \times 10^{-3} = \boxed{758\ kN}$
	<u>Stiff bearing factor C_2</u>
	For 1mm <u>length</u> of stiff bearing :
	area at end of fillet $= 1 \times 18{\cdot}6 = 18{\cdot}6\ mm^2$
	$C_2 = 18{\cdot}6 \times 340 \times 10^{-3} = \boxed{6{\cdot}32\ kN/mm}$
	<u>Flange plate factor C_3</u>
	For 1mm <u>thickness</u> of flange plate :
	area at end of fillet $= 2{\cdot}5 \times 18{\cdot}6 = 46{\cdot}5\ mm^2$
	$C_3 = 46{\cdot}5 \times 340 \times 10^{-3} = \boxed{15{\cdot}8\ kN/mm}$

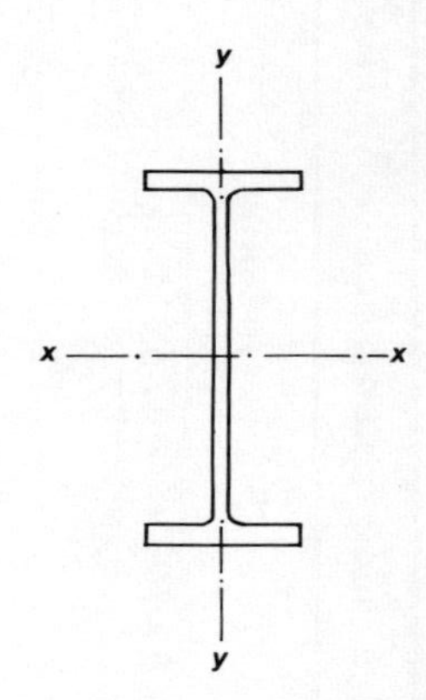

Buckling resistance moment for steel grade 50

Designation serial size mass/metre and capacity		Buckling resistance moment M_b (kNm) for effective length (L_E) in metres with variable slenderness correction factors (n)												
	n	4.0	5.0	6.0	7.0	8.0	9.0	10.0	11.0	12.0	13.0	14.0	15.0	16.0
610 × 305 × 238	0.4	2540	2540	2540	2540	2520	2460	2410	2360	2310	2260	2210	2170	2120
M_{cx} = 2540	0.6	2540	2490	2380	2270	2170	2070	1970	1890	1800	1730	1650	1590	1520
Plastic	0.8	2420	2240	2080	1920	1780	1640	1520	1420	1330	1240	1170	1100	1040
	1.0	2210	1980	1760	1570	1400	1270	1150	1050	968	897	836	783	736

M_b is obtained using an equivalent slenderness = n.u.v. L_E/ry
Values have not been given for values of slenderness greater than 300
The section classification given applies to members subject to bending only.

Figure 3.15 Buckling resistance moments for universal beams. (source: Steel Construction Institute)

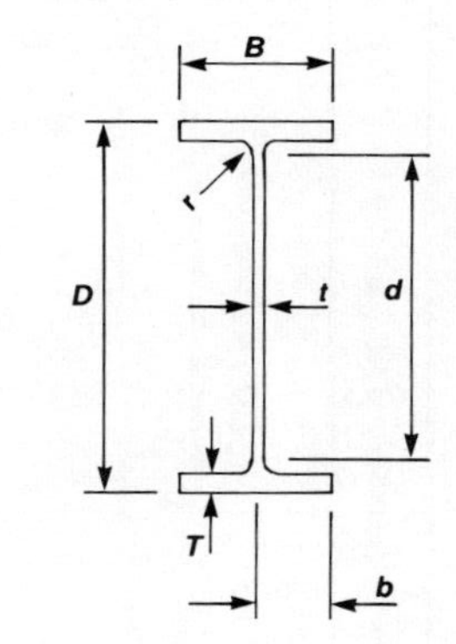

Bearing and buckling values for unstiffened webs for steel grade 50

Designation serial size and mass/metre (mm)	(kg)	Web thickness t (mm)	Depth between fillets d (mm)	Bearing – End bearing – Beam factor C1 (kN)	Bearing – End bearing – Stiff bearing factor C2 (kN/mm)	Bearing – End bearing – Flange plate factor C3 (kN/mm)	Bearing – Continuous over bearing – Beam factor C1 (kN)	Bearing – Continuous over bearing – Stiff bearing factor C2 (kN/mm)	Bearing – Continuous over bearing – Flange plate factor C3 (kN/mm)	Buckling – End bearing – Beam factor C1 (kN)	Buckling – End bearing – Stiff bearing factor C2 (kN/mm)	Buckling – End bearing – Flange plate factor C3 (kN/mm)	Buckling – Continuous over bearing – Beam factor C1 (kN)	Buckling – Continuous over bearing – Stiff bearing factor C2 (kN/mm)	Buckling – Continuous over bearing – Flange plate factor C3 (kN/mm)	Shear value P_v (kN)
610 × 305	238	18.6	537	757	6.32	15.8	1510	6.32	31.6	1210	3.81	3.81	2410	3.81	7.62	2400
	179	14.1	537	481	4.79	12.0	962	4.79	24.0	643	2.08	2.08	1290	2.08	4.17	1780
	149	11.9	537	366	4.05	10.1	732	4.05	20.2	418	1.37	1.37	836	1.37	2.74	1480

Web capacity = $C1 + b1.C2 + t_p \cdot C3$
where $b1$ is the length of stiff bearing and t_p is the thickness of flange plates or packing
Where the flange plate or packing is not continuously welded the factor C3, for bearing only, should be taken as equal to C3 divided by 2.5.

Figure 3.16 Web bearing, buckling and shear values for universal beams (source: Steel Construction Institute)

BS5950 clause	Example 3.7 Use of tables for beam design 3.7.1
	1500kN (factored)
	Continuous lateral support
	Grade 50 steel 12.0m
	Design moment $M = 1500 \times 12/8 = \boxed{2250 \text{ kNm}}$
	Design shear $F_v = 1500/2 = \boxed{750 \text{ kN}}$
	Try a 610 × 305 × 238 UB
	(see example 3.6.1 for strength etc)
	Values from Constrado Guide, vol. 1
	Moment capacity $M_{cx} = \boxed{2540 \text{ kNm}}$ (p257)
	> 2250 kNm
	Shear capacity $P_v = \boxed{2400 \text{kN}}$ (p 268)
	$0.6 P_v = 0.6 \times 2400 = 1440$ kN > 750 kN
	no reduction in moment capacity
	Web buckling beam factor $C_1 = 1210$ kN(p268)
	> 750 kN, nominal bearing required
	Web bearing beam factor $C_1 = 757$ kN (p 268)
	> 750kN
Table 5	Deflection allowed say span/360
	Imposed load to cause this deflection:
	$W = \left(\frac{384}{5} \quad \frac{12 \times 10^3}{360} \quad \frac{205 \times 10^3 \times 208000 \times 10^4}{12^3 \times 10^9} \right) \times 10^{-3}$
2.5.1	$= \boxed{632 \text{ kN}}$ (unfactored)

BS5950 clause	Example 3.7 Use of tables for beam design 3.7.2
	Redesign with effective lateral restraint at ends and midspan only
	$M = WL/8$
	$M_0 = WL/32$
	$\beta = 0$
	Beam loaded between adjacent restraints
Table 13	$m = 1.0$, n from Table 15 or 16
Table 16	$M/M_0 = \gamma = +4.0$
	$n \approx 0.87$
	From Constrado Guide, vol 1:
	for $L_E = 6.0\,m$, $n = 0.87$
	$M_b < 2080\,kNm < M$ (p257)
	thus the section is not adequate.
	However if lateral restraints are provided at 4.0m spacing
	$M_b > 2420\,kN\,m > M$

6.0 m
6.0 m
βM
M
M_0

shortened by using approximations for u (0.9 for rolled sections, 1.0 for others), v (generally 1.0) and n (1.0).

The effective length L_E is affected by the restraint afforded to the beam and on whether the load does or does not have a destabilising effect (Figure 3.14). The basic case in which L_E is equal to the actual length between supports occurs for a beam loaded normally (i.e. without destabilising load) under the following conditions at the supports:

(a) torsional restraint
(b) compression flange laterally restrained
(c) compression flange only free to rotate on plan.

3.6.2 *Rolled sections used as beams*

The design of rolled section beams is facilitated by the use of tables (3). From these the buckling resistance moment and web shear, bearing and buckling values may be obtained so enabling a suitable section to be selected with a minimum of calculation.

Figure 3.15 shows an extract from the buckling resistance moment table for universal beams in grade 50 steel, giving the values of M_b for a range of effective lengths L_E and slenderness correction factor n. The values of M_b are calculated from precise values of u and n. The section classification and the maximum moment capacity M_{cx} are also tabulated. Figure 3.16 shows an extract from the table of bearing, buckling and shear values for unstiffened webs in grade 50 steel. Design of a rolled beam thus only requires a check on the bending and web capacities and a final estimate of deflection as illustrated in Example 3.7.

3.7 Compound beams

Where there is some restriction on the depth of a rolled beam, for example for architectural reasons, and where the largest rolled beam within the allowable depth has insufficient strength it may be reinforced by welding plates to the flanges, forming a compound beam.

The area of flange plate required may be determined from the difference between the plastic section modulus required and that actually provided by the rolled beam:

$$S_{req} = S_{beam} + S_{plates}$$

If the distance between the flanges is taken as D (Figure 3.17), the beam depth, then:

$$S_{plates} = A_{plate}D$$

from which the area of each plate:

$$A_{plate} = (S_{req} - S_{beam})/D.$$

This is in fact somewhat larger than the area actually needed.

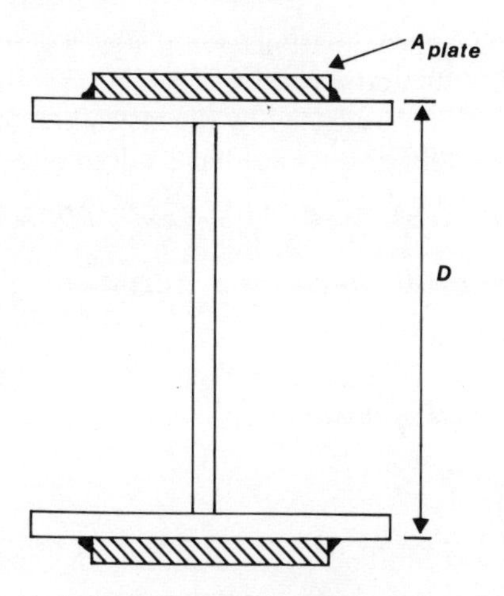

Figure 3.17 Compound beam

It must be borne in mind that the added plate elements will themselves contribute to the classification of the compound beam for local buckling purposes. It is also necessary to recognise that if plates are curtailed the beam no longer has a constant cross-section throughout its length, a fact which will affect the value of slenderness correction factor n and equivalent uniform moment factor m (see Example 3.8).

3.8 Plate girders

The range of application of rolled beams is wide; nevertheless there are limits to their employment dictated by loading, span or other considerations. Outside these limits the designer is free to devise a suitable beam from first principles—one popular solution is a welded plate girder which consists simply of two equal or unequal size flange plates welded to a web plate (Figure 3.18).

Optimum depths and steel areas can be established from the formulae given earlier once a decision has been made on the web depth to thickness ratio. Generally the more slender the web the more economic the girder in terms of steel weight. However it is not possible to be dogmatic on this point as much will depend on the magnitude of loading. In some cases also the optimum depth may be unacceptably large; a shallower web will be used, leading in turn to a less efficient girder.

High bending efficiency implies flanges that are separated by the thinnest practicable web. Thin webs, in turn, imply some stiffening system to give stability. Economy is, however, more than just an attempt to use the minimum amount of material: fabrication costs can erode material cost savings considerably or even outweigh them.

Plate girder design, in which relatively slender elements are to be used, is mainly concerned with the local stability of webs and flanges and the provision

BS5950 clause	Example 3.8 Compound beam 3.8.1
	Design data
	Total depth not to exceed 600 mm
	Dead load (including allowance for self weight) 17.5 kN/m
	Imposed load 22.0 kN/m
	Span 12.0 m
	Full compression flange restraint
	Grade 43 steel
	Loading
2.4	Ultimate limit state
	Dead load 17.5 × 1.4 = 24.5 kN/m
	Imposed load 22.0 × 1.6 = 35.2 kN/m
	Total load 59.7 kN/m
	M_{max} 0.125 × 59.7 × 12[3] = **1075 kNm**
	F_{max} 59.7 × 12/2 = **358 kN**
	Largest available universal beam within the depth limitation:
	533 × 210 × 122 UB
	D 544.6 mm, t 12.8 mm
	B 211.9 mm, T 21.3 mm
	S 3200 cm³
Table 6	Steel 16 < T < 40 mm, p_y = 265 N/mm²

BS5950 clause	Example 3.8 Compound beam 3.8.2
	Moment capacity
	$M_{cx} = 265 \times 3200 \times 10^{-3}$ $= 848$ kNm
	Flange plates must provide:
	$1075 - 848$ $= 227$ kNm
	$S_{req} = 227 \times 10^6 / 265$ $= 857 \times 10^3$ mm^3
	$A_{plate} \approx 857 \times 10^3 / 544.6$ $= 1570$ mm^2
3.5.2	For section to be compact:
Table 7	plate $b/T \leqslant 8.5\varepsilon$
	$\varepsilon = \left(\frac{275}{265}\right)^{1/2}$ $= 1.02$
	try plates 15×120 $= 1800$ mm^2
3.5.5 (b)	$b/T = 120/15 = 8$ $< 8.5 \times 1.02$
	$S_{plates} = 120 \times 15 \times (544.6 + 15)$ $= 1.01 \times 10^6$ mm^3
	$> 857 \times 10^3$ mm^3
	Adopt 15 × 120 flange plates
	Cut-off point
	Plates may be discontinued where the bending moment falls below 848 kNm. From the geometry of the bending moment diagram this occurs at:
	$\left\{\frac{(1075 - 848)}{1075} \times 6^2\right\}^{1/2}$ $= 2.75$ m
	either side of mid span.

BS5950 clause	Example 3.8 Compound beam 3.8.3
	BMD (kNm): 848, 1075, 848
	2.75m, 2.75m; 6.0m, 6.0m
	Weld at end of flange plates
6.1.1	Design to transmit force in flange plate at cut-off point
	Total S = 3200 + 1010 = 4210 cm^3
	Stress in plate = $848 \times 10^6 / 4210 \times 10^3$ = 201 N/mm^2
	force in plate = $15 \times 120 \times 201 \times 10^{-3}$ = 362 kN
6.6.5.1 Table 36	Fillet weld design strength 215 N/mm^2
	length of 6mm fillet weld required:
6.6.5.3	$(362 \times 10^3)/215 \times 6 \times 0.7$ = 401 mm
6.6.5.2	total length 401 + 2 x 6 = 413 mm
	extension; theoretical cut-off point; 120
	extension = 0.5 (413 − 120) = 146 mm

BS5950 clause	Example 3.8 Compound beam	3.8.4

Welding plates to flanges

Vertical shear at cut-off point:

$358 \times 2.75/6.0 = 164$ kN

$A = 15 \times 120 = 1800\ mm^2$

$\bar{y} = (544.6 + 15)/2 = 279.8$ mm

$I_{beam} = 762 \times 10^6\ mm^4$

$I_{plates} = 2 \times (15 \times 120) \times 279.8^2 = 282 \times 10^6\ mm^4$

$I_{total} = (762 + 282) \times 10^6 = 1.04 \times 10^9\ mm^4$

Horizontal shear between plate and beam flange:

$$\frac{164 \times 10^3 \times 1800 \times 279.8}{1.04 \times 10^9} = 79.4\ N/mm$$

Load capacity of 6mm fillet weld:

$6 \times 0.7 \times 215 = 903$ N/mm

ratio weld/(weld + space):

$79.4/(2 \times 903) = 0.044$

6.6.2.5 — Maximum spacing

(i) in no case more than 300 mm

(ii) 16 times thickness of thinner part joined in compression

BS5950 clause	Example 3.8 Compound beam 3.8.5
	(iii) 24 times thickness of thinner part joined in tension
	Thinner part is flange plate (15 < 21·3)
	16 × 15 = 240 mm < 300 ; 240 mm rules.
	a effective length of weld
	b actual length of weld
	c spacing between effective lengths
	a/(a+c) = 0·044
	for c = 240 mm
	(1 − 0·044)a = 0·044 × 240 gives a = 11 mm
	b = 11 + (2 × 6) = 23 mm
	ie intermittent welding can be used.
	23 mm is too short for efficient welding − adopt 50 mm weld lengths spaced at 240 mm

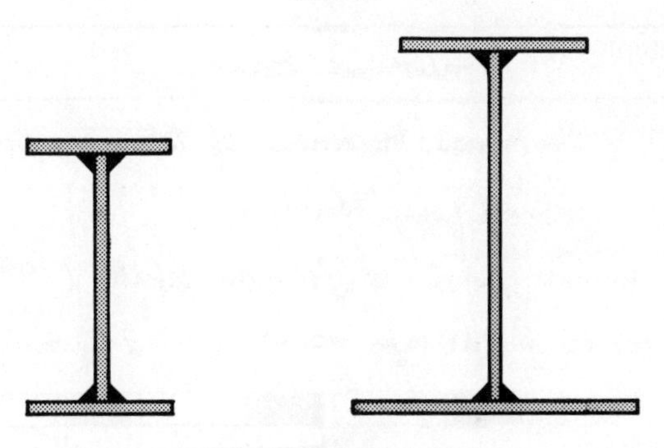

Figure 3.18 Plate girders

of suitable stiffening to the former, to inhibit shear buckling, permit tension field action and cope with the effects of concentrated loads and reactions.

3.8.1 *Bending*

In general the moment capacity of a plate girder is governed by exactly the same considerations as any other beam, that is to say its capacity depends on the section classification (plastic, compact, semi-compact or slender) modified as necessary for high shear load. However, plate girders will, for efficiency, have *thin webs* ($d/t > 63\varepsilon$) and for such special considerations apply. The moment capacity for sections with plastic, compact or semi-compact flanges and with thin webs may be assumed to be provided by the flanges alone, the web being considered solely as a shear carrying element. For such a section with plastic or compact flanges $M_c = p_{yf}S_{xf}$, where p_{yf} is the design strength of the flange material and S_{xf} is the plastic modulus of the flanges only, about the section axis.

Plate girders with slender flanges are uncommon in building structures but if used their moment capacity is calculated from the reduced stress.

3.8.2 *Shear*

It has been noted (Section 3.4) that the shear capacity of a thin web can be increased by appropriate vertical stiffening and that further capacity is available in basic tension field action provided that the girder end panels are strong enough to anchor the forces induced by the tension field. Additional tension field strength is available from a flange dependent contribution if the flanges are of sufficient rigidity. The maximum possible strength of a web is 0.6 of the design strength.

For economy consideration will need to be given to the conflict between using a heavily stiffened thin web or less stiffening on a thinner web. Figure 3.19 provides some guidance in showing how much very thin webs benefit from a reduction in stiffener spacing. See also Examples 3.9 and 3.10.

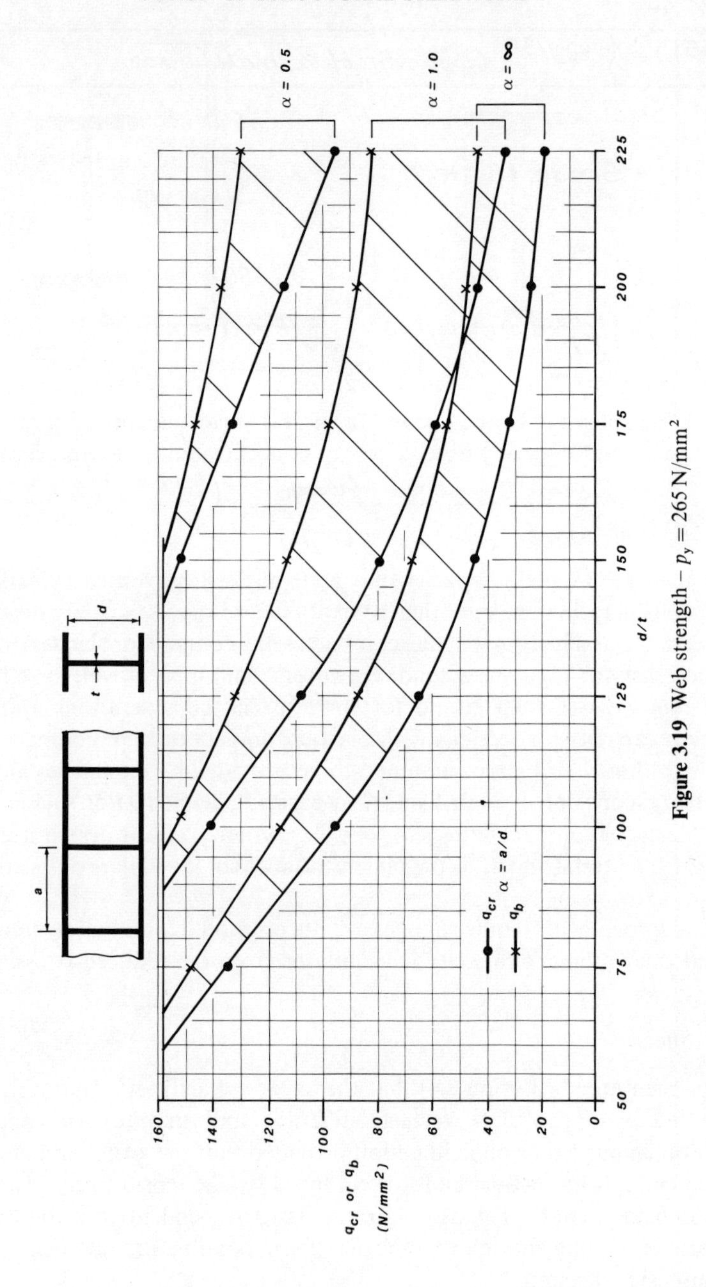

Figure 3.19 Web strength – $p_y = 265\ \text{N/mm}^2$

BS5950 clause	Example 3.9 *Capacity of a plate girder* 3.9.1
	Grade 43 steel
	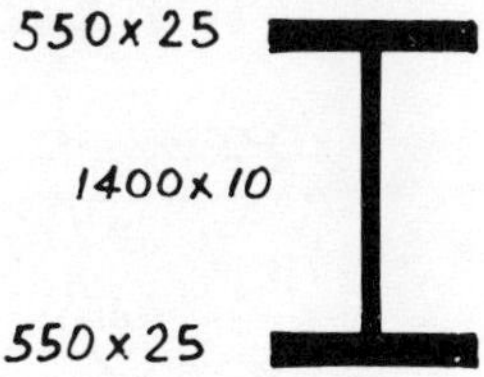
3.1.1 Table 6	$p_y = 265\ N/mm^2 \quad (16 < T < 40\ mm)$
	$b/T = (550-10)/(2\times25) = 10.8$
	$\varepsilon = (275/265)^{1/2} = 1.02$
3.5.2 Table 7	semi compact flange $(b/T < 13\varepsilon)$
	web $d/t = 1400/10 = 140$
note 2	thin web : $140 > 63\varepsilon$
	Moment capacity:
4.4.4.2(a)	flanges resist moment
	$S_{fx} = 550\times25\times(1400-25) = 18.9\times10^6\ mm^3$
	$M_c = 265\times18.9\times10^6\times10^{-6} =$ **5010 kN m**
	Shear buckling resistance:
4.4.5.1	Shear carried by web:
	a. Without tension field action
4.4.5.3	unstiffened web $a/d = \infty$
Table 21a	$q_{cr} = 51\ N/mm^2$
	$V_{cr} = 51\times1400\times10\times10^{-3} =$ **714 kN**
	b. With tension field action
	stiffened web a/d say 1.0
Table 22a	$q_b = 118\ N/mm^2$

BS5950 clause	Example 3.9 Capacity of a plate girder 3.9.2
4.4.5.4.1	$V_b = 118 \times 1400 \times 10 \times 10^{-3} = \boxed{1650\ kN}$
	<u>Increase if flanges not fully stressed:</u>
Table 23a	$q_f = 307\ N/mm^2$
	$K_f = \frac{M_{pf}}{4 M_{pw}} \left(1 - \frac{f}{p_{yf}}\right)$
	$M_{pf} = \frac{550 \times 25^2}{4} p_y = 85.9 \times 10^3\ p_y\ Nmm$
	$M_{pw} = \frac{10 \times 1400^2}{4} p_y = 4.9 \times 10^6\ p_y\ Nmm$
	$\frac{f}{p_{yf}}$ say 0.75
	$K_f = \frac{85.9 \times 10^3}{4 \times 4.9 \times 10^6} (1 - 0.75) = 1.1 \times 10^{-3}$
	$q_f (K_f)^{1/2} = 307 \times (1.1 \times 10^{-3})^{1/2} = 10.2\ N/mm^2$
	$V_b = (118 + 10.2) \times 1400 \times 10 \times 10^{-3} = 1800\ kN$
4.4.5.4.1	$0.6\ p_y\ dt = 0.6 \times 265 \times 1400 \times 10 \times 10^{-3} = 2230\ kN$
	$\therefore$ maximum $V_b = \boxed{1800\ kN}$

BS5950 clause	Example 3.10 Design of a welded plate girder 3.10.1
	On the ground floor of a multi-storey building having a column spacing of 4.0 m it is required to have an unobstructed span of 16.0 m. A plate girder is to be designed to carry the column loads from the floors above

4.0m 4.0m 4.0m
800 800 800 kN 30kN/m
16.0m

FACTORED LOADS

5520 7360

FACTORED BM kNm

1440 1320 400

FACTORED SHEAR kN

BS5950 clause	Example 3.10 Design of a welded plate girder 3.10.2
	Steel grade 50
	assume thickness > 16 mm but < 40 mm
Table 6	$p_y = 340\ N/mm^2$
	assume web slenderness, $d/t = 140$
	optimum girder size (see chapter 3.3)
	$S = 7360 \times 10^6/340 = \boxed{21.7 \times 10^6\ mm^3}$
	optimum depth $h = 1.26\,(140 \times 21.7 \times 10^6)^{1/3}$
	$= \boxed{1.83\ m}$
	optimum area $A = 2.38\left[(21.7 \times 10^6)^2/140\right]^{1/3}$
	$= \boxed{35.7 \times 10^3\ mm^2}$
	Trial girder
	For architectural reasons a depth of 1.83 m is excessive – adopt $d = 1.50\ m$.
	web thickness $t = 1500/140$ – say 10 mm
	flange area $(7360 \times 10^6/1500 \times 340) = 14.4 \times 10^3\ mm^2$
	try flanges 550 x 30 $= 16.5 \times 10^3\ mm^2$
	Section classification
Fig. 3	flange $b = (550-10)/2 = 270\ mm$
	$b/T = 270/30 = 9.0$
	$\varepsilon = (275/340)^{1/2} = 0.899$
	550x30 – 1500x10 – 550x30 –
Table 7	limiting values of b/T:
	compact $8.5 \times 0.899 = 7.64$
	semi-compact $13 \times 0.899 = 11.7$

BS5950 clause	Example 3.10 Design of a welded plate girder 3.10.3
	flange is semi-compact
	web $d/t = 1500/10 = 150 > 63\varepsilon$
Table 7 note 2	check web for shear buckling
4.2.3	Check shear
(4.4.1)	$F_v = $ 1440 kN
	$A_v = 1500 \times 10 = 15.0 \times 10^3\ mm^2$
	$P_v = 0.6 \times 340 \times 15.0 \times 10^3 \times 10^{-3} = $ 3060 kN
	$P_v > F_v$
4.4.2	Minimum web thickness
	Assume transverse stiffeners at spacing $a > d$
4.4.2.2	for serviceability $t \geq d/250$
(b)(1)	$d/250 = 1500/250 = 6.0\ mm < 10.0\ mm$
4.4.2.3	to avoid flange buckling
(b)(2)	assume $a \leq 1.5d$; $t \geq \left(\frac{d}{250}\frac{p_{yf}}{455}\right)^{1/2}$
	$t \geq \left(\frac{1500}{250}\frac{340}{455}\right)^{1/2} = 5.2\ mm$
	10 mm web suitable
4.4.4	Moment capacity
4.4.4.2	Moment resisted by flanges:
(a)	$M_c = 340 \times 16.5 \times 10^3 \times 1530 \times 10^{-6} = $ 8580 kNm
	> 7360 kNm
4.3	Lateral torsional buckling
(4.4.1)	Compression flange continuously restrained

BS5950 clause	Example 3.10 Design of a welded plate girder 3.10.4
4.4.5	Web
4.4.5.3	(a) Design without using tension field action -
	at ends $q_{cr} \geq 1440 \times 10^3/1500 \times 10 = 96\ N/mm^2$
Table 21(c)	for spacing $a/d = 0.8$, q_{cr} = 103 N/mm²
	$a = 0.8 \times 1500 = 1200$ mm
	spacing can be increased towards mid span in accordance with the shear force diagram.
	(b) Using tension field action, giving a more economical stiffener arrangement.
	for spacing $a/d = 1.6$ and $d/t = 150$
Table 22(c)	q_b = 98 N/mm²
	800 kN, 800 kN; 2.4 m, 1.6 m, 0.8 m, 3.2 m; 1.5 m; 1440 A, 1368 B, 1320 C, 496 D, 400 E shear kN
	Proposed stiffener layout
4.4.5	Shear buckling resistance:
	Panel A-B
4.4.5.2	Design using tension field action
	q_b for $a/d = 2.4/1.5 = 1.6$:-
Table 22(c)	$q_b = 98\ N/mm^2$

BS5950 clause	Example 3.10 Design of a welded plate girder 3.10.5
4.4.5.4.1	$V_b = 98 \times 1500 \times 10 \times 10^{-3} = \boxed{1470\ kN} > 1440\ kN$
	<u>Panel B-D</u>
	By inspection adequate.
	<u>Panel D-E</u>
	$a/d = 3{\cdot}2/1{\cdot}5 = 2{\cdot}13$
Table 22(c)	$q_b = 85\ N/mm^2$
	$V_b = 85 \times 1500 \times 10 \times 10^{-3} = 1280\ kN > 496\ kN$
	[stiffeners B and D are intermediate and do not carry imposed load, stiffeners A, C and E are load bearing]
4.4.6	<u>Design of intermediate stiffeners.</u>
4.4.6.1	Stiffeners both sides of web
4.4.6.2	Spacing - see diagram p. 3.10.4
4.4.6.3	Outstand
	try stiffener 100 × 8 mm
4.5.1.2	$19 t_s \varepsilon = 19 \times 8 \times 0{\cdot}899 = 137 > 100\ mm$
	$13 t_s \varepsilon = 13 \times 8 \times 0{\cdot}899 = 93{\cdot}5 < 100\ mm$
	core section is 93·5 × 8 mm
	minimum stiffness
	$a \geq \sqrt{2}\, d$

8

93·5 core

10

℄ web

BS5950 clause	Example 3.10 Design of a welded plate girder 3.10.6
	I_s =
	$2\times\left(\frac{8\times 93.5^3}{12}\right) = 1.09\times 10^6$
	$2\times 93.5\times 8\times 51.8^2 = 4.01\times 10^6$
	5.1×10^6 mm^4
	$0.75\,dt^3 = 0.75\times 1500\times 10^3 = 1.13\times 10^6$ mm^4
4.4.6.6	**Buckling check**
	core area; x — x; 200; 200
	I_{xx} = 5.1×10^6
	+ $400\times 10^3/12$ 0.033×10^6
	5.13×10^6 mm^4
	Area $2\times 93.5\times 8 = 1.50\times 10^3$
	$400\times 10 = 4.00\times 10^3$
	5.50×10^3 mm^2
4.5.1.5	$r = (5.13\times 10^6/5.50\times 10^3)^{1/2} = 30.5$ mm
4.5.1.5	$\lambda = 1500\times 0.7/30.5 = 34.4$
4.7.5	welded section - $p_y = 340-20 = 320$ N/mm^2
Table 27(c)	$p_c = 284$ N/mm^2
	$P_q = 5.50\times 10^3\times 284\times 10^{-3} = 1560$ kN
4.4.6.6	$F_q = V - V_s$
4.4.5.3	$V_s = q_{cr}\,dt$
Table 21(c)	for $a/d = 1.6$, $q_{cr} = 57$ N/mm^2

BS5950 clause	Example 3.10 Design of a welded plate girder 3.10.7
	$V_s = 57 \times 1500 \times 10 \times 10^{-3} = 855\text{ kN}$
	$F_q = 1368 - 855 = \boxed{513\text{ kN}} < 1540\text{ kN}$
4.4.6.7	Connection to web
	shear $t^2/8b_s = 10^2/8 \times 100 = 0.125\text{ kN/mm}^2$
	nominal welding required.
	Intermediate stiffeners 100×8 mm
4.4.5.4.3	End panel designed using tension field action
Figure 7	(double stiffener)
(b)	End post checked as a beam spanning between girder flanges:
4.4.5.4.4	forces on end post (anchor forces)
	$R_{tf} = \dfrac{H_q}{2}$
	$H_q = 0.75\, d\, t\, p_y \left(1 - \dfrac{q_{cr}}{0.6p_y}\right)^{1/2}$
	For panel A-B; $a/d = 1.6$, $d/t = 150$
Table 21(c)	$q_{cr} = 57\text{ N/mm}^2$
Table 22(c)	$q_b = 98\text{ N/mm}^2$
	$f_v = 1440 \times 10^3/1500 \times 10 = 96\text{ N/mm}^2$
	$H_q = 0.75 \times 1500 \times 10 \times 340\left(1 - \dfrac{57}{0.6 \times 340}\right)^{1/2} = 3.25 \times 10^6\text{ N} = 3250\text{ kN}$
4.4.5.4.4	$f_v < q_b$: reduce H_q by $\dfrac{f_v - q_{cr}}{q_b - q_{cr}}$

BS5950 clause	Example 3.10 Design of a welded plate girder 3.10.8
	$H_q = 3250 \times \left(\frac{96-57}{98-57}\right) = 3090$ kN $R_{tf} = 3090/2 = 1545$ kN End post horizontal web dimension h mm Shear capacity $0.6 \times 340 \times 10 \times h \times 10^{-3} = 2.04h$ kN $h \geq 1545/2.04 = 757$ mm make end post 800 mm long. $M_{tf} = H_q d/10 = 1545 \times 1.5/10 = 232$ kN m Compression in end stiffener - assuming distance between stiffeners is 800 mm: $232 \times 10^3/800 = 290$ kN capacity of 2/100x8 stiffeners: $2 \times 100 \times 8 \times 340 \times 10^{-3} = 544$ kN [Note that the end post increases the girder length by 0.8 m at each end. This dimension is dictated by the shear in the end post caused by tension field action. If such a projection is not acceptable the end post may be reduced in length, increasing q_{cr} and so reducing H_q]

BS5950 clause	Example 3.10 Design of a welded plate girder 3.10.9
4.5	Web bearing, buckling and stiffener design
	Stiffeners will be needed to strengthen the web against concentrated loading causing:
4.5.1.1(a)	(i) local buckling - load carrying
4.5.1.1(b)	(ii) local crushing - bearing
4.5.4	Load carrying stiffeners
	Worst case is stiffener at A
	external load F_x = 1440 kN
4.5.4.2	minimum area $0.8F_x/p_{ys}$ = $0.8 \times 1440 \times 10^3/340$
	= 3390 mm^2
	try 2 stiffeners 150x15mm = 4500 mm^2
4.5.4.1	Buckling check:
4.5.1.2	outstand within limits - use full area.
	stiffeners 150x15; 10; x; x; 200; 200
	I_{xx}
	$2 \times 150^3 \times 15/12$ = 8.44×10^6
	$2 \times 150 \times 15 \times 80^2$ = 28.80×10^6
	$400 \times 10^3/12$ = 0.033×10^6
	37.3×10^6 mm^4
	Area $2 \times 150 \times 15 + 400 \times 10$ = 8.5×10^3 mm^2
	$r = (37.3 \times 10^6/8.5 \times 10^3)^{1/2}$ = 66.2 mm
4.5.1.5(a)	$L_E = 0.7L$ (flange restrained)

BS5950 clause	Example 3,10 Design of a welded plate girder 3.10.10
4.5.1.5	$\lambda = 1500 \times 0.7/66.2 = 15.9$
4.7.5	welded section - $p_y = 340 - 20 = 320$ N/mm^2
Table 27(c)	$p_c = 318$ N/mm^2
	$P_z = 8.5 \times 10^3 \times 318 \times 10^{-3} = 2700$ kN $> F_x$
4.4.6.6	Buckling check for combined action
	$F_q = 1440 - 855 = 585$ kN
	$F_q < F_z$
	stiffener meets buckling check.
4.5.3	Bearing check
4.5.1.3	assume stiff bearing length $b_1 = 0$
4.5.3	local capacity of web
	$n_1/2$ $n_1/2$ 30 $\tan^{-1} 1/2.5$ A
	$n_1 = 2 \times 30 \times 2.5 = 150$ mm
	local capacity of web $= 150 \times 10 \times 340 \times 10^{-3}$
	$= 510$ kN
	bearing load $1440 - 510 = 930$ kN
	stiffener capacity $8.5 \times 10^3 \times 340 \times 10^{-3}$
	$= 2890$ kN
	Adopt 2/150x15 mm stiffeners at A

BS5950 clause	Example 3.10 Design of a welded plate girder 3.10.11
	Similar checks should be carried out on the load carrying and bearing stiffeners at C and E.
4.5.2.2	Web check between stiffeners
	largest panel length D-E = 3·2 m
	load intensity 30 kN/m = 30 N/mm
(c)	f_{ed} = 30/10 = 3 N/mm²
	flange restrained against rotation relative to the web
	$p_{ed} = \left[2.75 + \frac{2}{(a/d)^2}\right] \frac{E}{(d/t)^2}$
	$= \left[2.75 + \frac{2}{\left(\frac{3200}{1500}\right)^2}\right] \frac{20500}{150^2} = 29.1\ N/mm^2$
	$p_{ed} > f_{ed}$ ∴ 10 mm thick web adequate.

3.9 Miscellaneous beams

3.9.1 *Gantry girders*

A moving overhead gantry crane is a common piece of equipment in industrial buildings. It runs between rails supported by girders which are in turn supported on columns. When the crane is stopped or started, or if the load accidentally slips, impact and surge loads occur. Allowance is made for these effects by increasing the static loadings by impact and surge factors. Additionally as the crane travels along the rails it tends to rotate (crab) about a vertical axis imposing a horizontal couple on the rails.

By the nature of their use gantry girders can not generally be provided with lateral restraint. The wheel loads act on the top flange and as both are free to move laterally the load must be considered a destabilising one. In order to strengthen the girder against lateral forces the top flange is reinforced by an additional element welded to it; commonly a channel or plate. See Example 3.11.

3.9.2 *Composite beams* (4)

Where steel beams support concrete slabs a reduction in steel beam weight is possible by interconnecting beam and slab. The resulting two-material beam is known as a composite beam. For full treatment of the fundamentals the reader is referred to a specialist text; the example which follows gives an indication of the method of ultimate load design with a serviceability check. Part 3 of British Standard 5950, Code of practice for Design in Composite Construction, is in preparation.

The elastic and plastic section properties of a composite beam may be calculated as shown in Example 3.12. Data are required concerning the material properties of the concrete; in particular its reaction to short and long term loading and its average stress at failure. Two moduluses of elasticity are commonly used to take account of the duration of loading. For ultimate load conditions the failure stress in concrete is assumed to be a proportion of the cube strength and to be uniformly distributed. Design tables are available (5).

A complication which arises where a number of steel beams, relatively widely spaced, are composite with a concrete slab is that the elementary theory of beam bending, which assumes a constant stress across a beam at any horizontal level, may not be valid. The width of slab associated with any one steel beam may then have to be curtailed to allow for the non-uniform stress distribution. The theoretical problem is complicated; in practice the reduced width (effective width) is found from simple formulae of the sort shown in Example 3.13.

3.10 Plastic design (6)

Restricting consideration to plastic collapse and noting that there is no economy in the plastic design of statically determinate beams the design

BS5950 clause	Example 3.11 Gantry girder	3.11.1

4.11

1.0 min

crab weight 25 kN

R_1

Hook load 100kN

Bridge weight 100kN

15.0 m

$R_1 = (100/2) + (100+25)(14/15) = 167\ kN$

R_1

1.5 1.5

crane girder span 8.0m

static wheel load = 167/2 = 83.5 kN

2.2.3 add 30% impact allowance and factor:

2.4.1.2 maximum vertical wheel load:

$83.5 \times 1.3 \times 1.6 = \boxed{174\ kN}$

4.11.2 Assume 10% transverse surge has greater effect than crabbing force

factored horizontal wheel load

$0.1 \times (100+25) \times 1.6/2 = \boxed{10kN}$

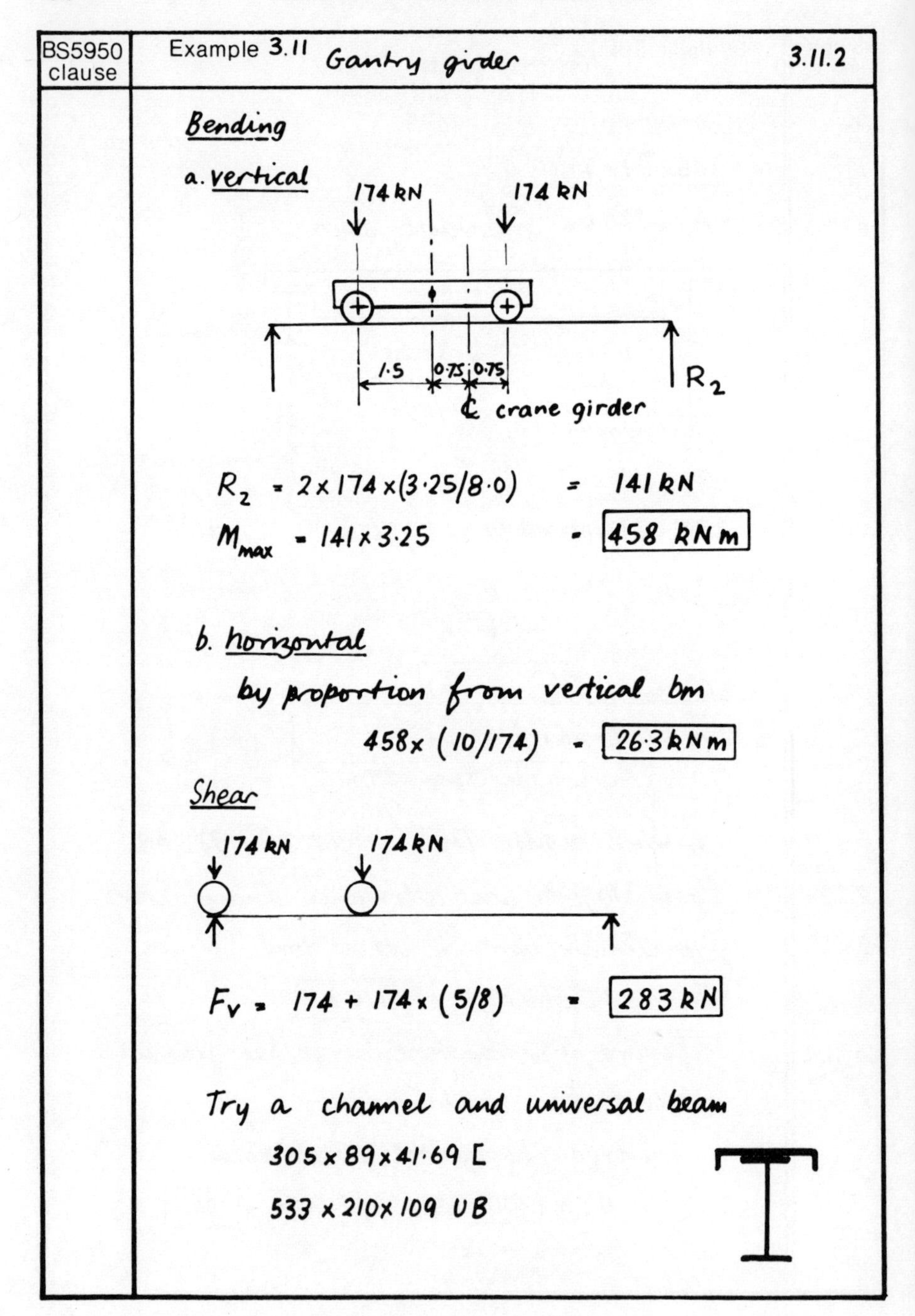

BS5950 clause	Example 3.11 Gantry girder	3.11.2

Bending

a. vertical

$R_2 = 2 \times 174 \times (3{\cdot}25/8{\cdot}0) = 141$ kN

$M_{max} = 141 \times 3{\cdot}25 =$ **458 kNm**

b. horizontal

by proportion from vertical bm

$458 \times (10/174) =$ **26·3 kNm**

Shear

$F_v = 174 + 174 \times (5/8) =$ **283 kN**

Try a channel and universal beam

305 × 89 × 41·69 [

533 × 210 × 109 UB

BS5950 clause	Example 3.11 Gantry girder 3.11.3

Components

305 × 89 × 41.69 [

A 53.1 cm²

I_{xx} 7060 cm⁴

I_{yy} 325 cm⁴

533 × 210 × 109 UB

A 139 cm²

I_{xx} 66700 cm⁴

I_{yy} 2940 cm⁴

Crane girder

$A = 139 + 53.1 = 192\ cm^2$

$$y_t = \left\{139 \times \frac{539.5}{2} + 53.1\,(539.5 + 10.2 - 21.8)\right\} / 192$$

$= 341$ mm

$y_c = 539.5 + 10.2 - 341 = 209$ mm

$$I_x = 66700 + 325 = 67025$$

$$+ 139\left(34.1 - \frac{53.9}{2}\right)^2 = 7106$$

$$+ 53.1(53.9 + 1.02 - 2.18 - 34.1)^2 = 18450$$

$$I_{xx} = 92600\ cm^4$$

BS5950 clause	Example 3.11 *Gantry girder* 3.11.4

$I_{yy} = 2940 + 7060 = 10000\ cm^4$

$r_{yy} = (10000/192)^{1/2} = 7.22\ cm$

top flange $I_{cf} = 7060 + \frac{1.88 \times 21.07^3}{12} = 8530\ cm^4$

bottom flange $I_{tf} = \frac{1.88 \times 21.07^3}{12} = 1470\ cm^4$

To calculate further section properties assume beam and channel to be composed of rectangular plates. An initial trial will show that the equal area axis (plastic neutral axis) cuts the flanges of the channel:

① + ② + ③ + ④ = A/2

① $(305 - 2 \times 13.7) \times 10.2 = 2.83 \times 10^3$

② $2 \times 13.7 \times x = 27.4x$

③ $210.7 \times 18.8 = 3.96 \times 10^3$

④ $(x - 10.2 - 18.8) \times 11.6 = 11.6x - 336$

$39x + 6.45 \times 10^3\ mm^2$

Total area $= 192\ cm^2 = 19.2 \times 10^3\ mm^2$

$39x + 6.45 \times 10^3 = 19.2 \times 10^3/2$

$x = 80.8\ mm$

$y = (539.5/2) + 10.2 - 80.8 = 199\ mm$

BS5950 clause	Example 3.11 Gantry girder 3.11.5

Calculate first moment of area

	area mm^2	$\bar{y}$ mm	$A\bar{y}$ mm^3
①	2.83×10^3	75.7	214×10^3
②	2.21×10^3	40.4	89.3×10^3
③	3.96×10^3	61.2	242×10^3
④	0.601×10^3	25.9	15.6×10^3
⑤	0.222×10^3	4.05	889
⑥	5.22×10^3	225	1.18×10^6
⑦	3.96×10^3	460	1.82×10^6
			3.56×10^6

$$S_x = 3.56\times10^6\,mm^3 = 3560\,cm^3$$

Section is symmetrical about minor axis

B.2.5.1(b)

$$u = \left(\frac{4S_x^2\gamma}{A^2h_s^2}\right)^{1/4}$$

$$\gamma = 1-\frac{I_y}{I_x} = 1-\frac{10000}{92600} = 0.892$$

$$u = \left(\frac{4\times3560^2\times10^6\times0.892}{192^2\times10^4\times520.7^2}\right)^{1/4} = 0.820$$

[h_s taken, approximately, as distance between flange centroids]

BS5950 clause	Example 3.11 Gantry girder 3.11.6
B.2.5.1(c)	$x = h_s \left(\frac{\Sigma bt + h_w t_w}{\Sigma bt^3 + h_w t_w^3} \right)^{1/2} = h_s \left(\frac{A}{\Sigma bt^3 + h_w t_w^3} \right)^{1/2}$
	$\Sigma bt^3 + h_w t_w^3$
	$277.6 \times 10.2^3 = 295 \times 10^3$
	$2 \times 210.7 \times 18.8^3 = 2.8 \times 10^6$
	$2 \times 89 \times 13.7^3 = 458 \times 10^3$
	$501.9 \times 11.6^3 = 783 \times 10^3$
	4.34×10^6
	$x = 520.7 \left(\frac{19200}{4.34 \times 10^6} \right)^{1/2} = 34.6$
4.3.5	<u>Effective length</u>
4.3.4	destabilising load
	compression flange laterally unrestrained and free to rotate on plan.
Table 9	$L_E = 1.2\,(L + 2D)$
	$= 1.2\,(8.0 + 2 \times 0.550) = 10.9\,m$
	$\lambda = 10.9 / 7.22 \times 10^{-2} = 151$
	$\lambda/x = 151/34.6 = 4.36$
B.2.5.1(d)	Slenderness factor v
	$N = 8530/(8530 + 1470) = 0.853$
Table 14 note 2a	Section has lipped flange

BS5950 clause	Example 3.11 Gantry girder 3.11.7
	Monosymmetry index ψ
	$D_L = 88.9 - 10.2 = 78.7$
	$N > 0.5$
	$\psi = 0.8(2N-1)\left(1 + \frac{D_L}{2D}\right)$
	$= 0.8(2 \times 0.853 - 1)\left(1 + \frac{78.7}{2 \times 550}\right) = 0.605$
	$v = \left[\left(4 \times 0.853(1-0.853) + \frac{4.4^2}{20} + 0.605^2\right)^{1/2} + 0.605\right]^{-1/2}$
	$= 0.714$
4.3.7.5	**Equivalent slenderness** $(n = 1)$
	$\lambda_{LT} = 0.824 \times 0.714 \times 151 = 88.8$
	Design strength
Table 6	grade 43 steel, $16 < T < 40$, $p_y = 265$ N/mm²
4.3.7.4	**Bending strength**
Table 12	welded section $p_b = 132$ N/mm²
	Buckling resistance
4.3.7.3	$M_b = 3560 \times 132 \times 10^{-3} = \boxed{470\ kNm} > 458$ kNm
	For horizontal bending assume all load resisted by the channel
	$S = 557$ cm³ $Z = 463$ cm³
	$M_c = 265 \times 557 \times 10^{-3} = 148$ kNm
	or $1.2 \times 265 \times 463 \times 10^{-3} = \boxed{147\ kNm} > 26.3$ kNm

BS5950 clause	Example 3.11 Gantry girder	3.11.8
4.9	For combined bending	
Table 2	$M_x = 458 \times 1.4/1.6 \quad = 401 \text{ kNm}$	
	$M_y = 26.3 \times 1.4/1.6 \quad = 23 \text{ kNm}$	
4.8.3.3	$\frac{M_x}{M_b} = \frac{401}{470} \quad = 0.853$	
	$\frac{M_y}{p_y Z_y} = \frac{23}{265 \times 463 \times 10^{-3}} \quad = 0.188$	
	$\underline{1.04}$	
	(marginally > 1.0 - accept)	
4.2.3	Shear	
(b)	shear area $A_v = 11.6 \times 501.9 = 5.82 \times 10^3 \text{ mm}^2$	
	shear capacity $P_v = 0.6 \times 265 \times 5.82 \times 10^3 \times 10^{-3}$	
	$= \boxed{925 \text{ kN}} > 283 \text{ kN}$	
	Checks on bearing, buckling and deflection will also be required.	

BS5950 clause	Example 3.12 Composite beam properties 3.12.1
	effective width 1500; 125; 1; 2/3; 4
	$610 \times 305 \times 238$ UB
	$A = 304\ cm^2$
	$I = 208000\ cm^4$
	$D = 633.0\ mm$
	concrete strength $30\ N/mm^2$
Table 6	steel grade 43, $p_y = 265\ N/mm^2$
	Elastic
	modular ratio $E_{steel}/E_{concrete} = 15$
	transformed concrete area $1500 \times 125/15 = 12.5 \times 10^3\ mm^2$
	total area
	30.4×10^3
	12.5×10^3
	$42.9 \times 10^3\ mm^2$
	centroids: concrete, composite, steel; 62.5; 379.0; 316.5; d_c
	$d_c = \frac{30.4 \times 10^3 \times 379}{42.9 \times 10^3} = \boxed{269\ mm}$
	total second moment of area $I_t =$
	$(2.08 \times 10^9) + (30.4 \times 10^3 (379-269)^2) + \left(\frac{1500 \times 125^3}{12 \times 15}\right)$
	$+ (12.5 \times 10^3 \times 269^2)$
	$= \boxed{3.37 \times 10^9\ mm^4}$

BS5950 clause	Example 3.12 Composite beam properties 3.12.2

Section elastic moduluses

top of concrete $Z_1 = \dfrac{3{\cdot}37\times10^9}{269+62{\cdot}5} = \boxed{10{\cdot}2\times10^6\,mm^3}$

bottom of concrete $Z_2 = \dfrac{3{\cdot}37\times10^9}{269-62{\cdot}5} = \boxed{16{\cdot}3\times10^6\,mm^3}$

top of steel $Z_3 = \boxed{16{\cdot}3\times10^6\,mm^3}$

bottom of steel $Z_4 = \dfrac{3{\cdot}37\times10^9}{379+316{\cdot}5-269} = \boxed{7{\cdot}9\times10^6\,mm^3}$

[note that Z_1 and Z_2 are in equivalent steel units, and must be divided by the modular ratio when used to calculate the concrete stresses]

Ultimate

Ultimate stress in concrete say $0{\cdot}444\times30 = 13{\cdot}3\,N/mm^2$

maximum possible force in concrete:

$1500\times125\times13{\cdot}3\times10^{-3} = \boxed{2490\,kN}$

maximum possible force in steel:

$304\times10^2\times265\times10^{-3} = \boxed{8060\,kN}$

steel force > concrete force:

neutral axis must be in steel beam to increase compression

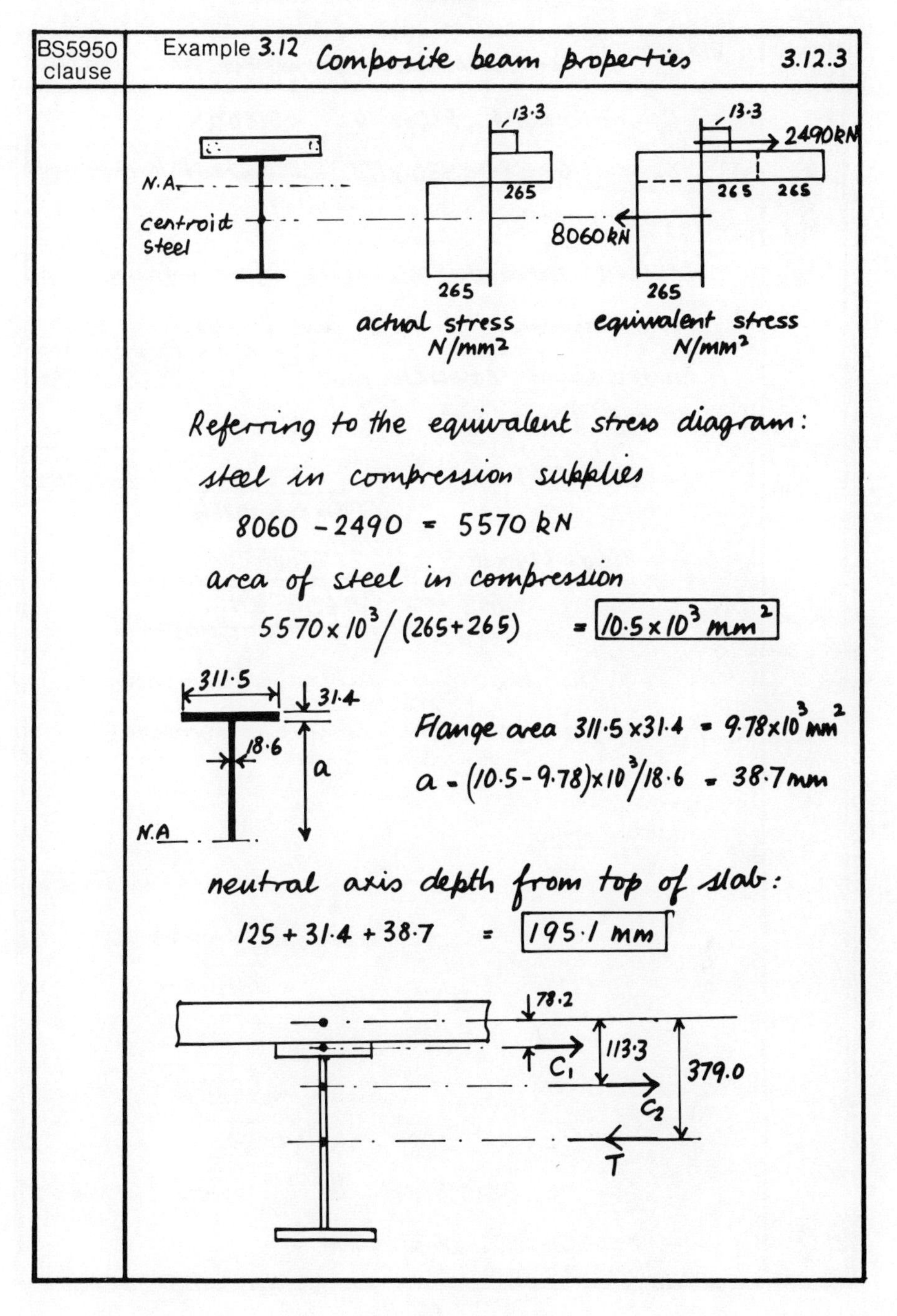
BS5950 clause
Example 3.12 Composite beam properties 3.12.3
13·3
13·3
2490kN
N.A.
265
265 265
centroid steel
8060kN
265
265
actual stress N/mm²
equivalent stress N/mm²
Referring to the equivalent stress diagram:
steel in compression supplies
8060 − 2490 = 5570 kN
area of steel in compression
5570 × 10³/(265 + 265) = 10·5 × 10³ mm²
311·5
31·4
18·6
a
N.A
Flange area 311·5 × 31·4 = 9·78 × 10³ mm²
a = (10·5 − 9·78) × 10³/18·6 = 38·7 mm
neutral axis depth from top of slab:
125 + 31·4 + 38·7 = 195·1 mm
78·2
113·3
379·0
C1
C2
T

BS5950 clause	Example 3.12 Composite beam properties 3.12.4
	$C_1 = 311{\cdot}5 \times 31{\cdot}4 \times 530 \times 10^{-3} = 5180\ \text{kN}$
	$C_2 = 18{\cdot}6 \times 38{\cdot}7 \times 530 \times 10^{-3} = 382\ \text{kN}$
	$T = 8060\ \text{kN}$
	To find ultimate moment of resistance take moments of C_1, C_2 and T about centroid of concrete slab:
	$5180 \times 78{\cdot}2 \times 10^{-3} = 405$
	$+\ 382 \times 113{\cdot}3 \times 10^{-3}$ $43{\cdot}3$
	$448{\cdot}3$ kNm
	$-\ 8060 \times 379 \times 10^{-3}$ $-\ 3055{\cdot}0$
	$M_u = \boxed{2606\ \text{kNm}}$

BS5950 clause	Example 3.13 Composite beam load capacity 3.13.1
	The universal beam of example 3.12 is used in a floor system, spaced at 3.0 m and spanning 12.0 m.
	Investigate the imposed loading which can be imposed on the beam:
	a. non composite with full lateral restraint
	b. composite - elastic design
	c. composite - ultimate load design.
	Materials as for example 3.12
	a. <u>Non composite.</u>
	<u>Dead load</u>:
	beam $238 \times 12 \times 9.81/10^3$ = 28 kN
	slab $0.125 \times 24 \times 3 \times 12$ = 108 kN
	136 kN
	factored dead load 136×1.4 = 190 kN
	For an unfactored total imposed load W_i kN : $M = (190 + 1.6W_i) \times 12/8 = (285 + 2.4W_i)$ kNm
	From Constrado manual vol 1, p 129:
	beam capacity M_{cx} = 1980 kNm
	$W_i = (1980 - 285)/2.4 = 706$ kN
Table 5	$\Delta = \dfrac{5 \times 706 \times 12^3 \times 10^9}{384 \times 205 \times 20800 \times 10^4} = 37.3\text{ mm} = \dfrac{\text{span}}{322}$
	imposed load/m² = $706/3 \times 12$ = **19.6 kN/m²**

BS5950 clause	Example 3.13 Composite beam load capacity 3.13.2
	b. Composite - elastic
	Steel beam carries all the dead load
	stress $= \frac{190 \times 12}{8} \times \frac{633 \times 10^6}{2 \times 208000 \times 10^4} = 43.4 N/mm^2$
	Top of slab and bottom of universal beam are critical stress levels:
	top of slab $1.6 W_i \times \frac{12}{8} \times \frac{1 \times 10^6}{10.2 \times 10^6 \times 15} = 13.3 N/mm^2$
	$W_i = 848 kN$
	bottom of beam $1.6 W_i \times \frac{12}{8} \times \frac{1 \times 10^6}{7.9 \times 10^6} = (265 - 43.4)$
	$W_i = 728 kN$
	bottom of beam rules
	$728/3 \times 12 = \boxed{20.2 kN/m^2}$
	[note the small increase in W_i over case a.; case b. assumes an elastic stress distribution but case a. utilises the full plastic moment of the universal beam]

BS5950 clause	Example 3.13 Composite beam load capacity 3.13.3
	c. Composite - ultimate load
	$1.6\ W_i \times \frac{12}{8} = 2606$ kNm
	$W_i = 1090$ kN
	$1090/3 \times 12 = $ **30.3 kN/m²**
	[note that the dead load stresses in the steel beam do not affect the ultimate moment of the composite beam]
	A note on the effective width of the concrete slab.
	A common rule for calculating the effective width is to take the least of :-
	i. ⅓ of the span 12/3 = 4.0m
	ii distance between beams = 3.0m
	iii 12 times slab depth 12×0.125 = 1.5m
	Thus effective width = 1.5m

BS5950 clause	Example 3.14 Plastic design of a 3 span beam 3.14.1
5.1	Factored imposed udl 50 kN/m
5.3	Factored dead udl 40 kN/m
5.4.2	A 7.0m B 10.0m C 9.0m D
	Free bending moments:
	AB $0.125 \times 90 \times 7^2$ = 551 kNm
	BC $0.125 \times 90 \times 10^2$ = 1130 kNm
	CD $0.125 \times 90 \times 9^2$ = 911 kNm
	Plastic moments:
	AB $90 \times 7^2/11.7$ = 377 kNm
	BC $90 \times 10^2/16$ = 563 kNm
	CD $90 \times 9^2/11.7$ = 623 kNm
	Assume full lateral restraint
	Steel grade 43, $p_y = 265\ N/mm^2$
	plastic modulus required for a uniform section, span CD critical:
	$S = 623 \times 10^3/265 \times 10^2 = 2350\ cm^3$
	533×210×101 UB $S = 2620\ cm^3$
	section is plastic: plastic design is permitted.

BS5950 clause	Example 3.14 Plastic design of a 3 span beam 3.14.2
	Check shear in span CD

4.2.3

$P_v = 0{\cdot}6\, p_y A_v$

$A_v = 10{\cdot}9 \times 536{\cdot}7 = 5{\cdot}85 \times 10^3\ \text{mm}^2$

$P_v = 0{\cdot}6 \times 265 \times 5{\cdot}85 =$ **930 kN**

F_v at C = **474 kN**

$0{\cdot}6\, P_v = 0{\cdot}6 \times 930 =$ **558 kN** $> F_v$

4.2.5

no reduction in moment capacity.

adopt 533×210×101 UB

[checks will also be needed on bearing, buckling and deflection]

problem is to provide a beam of sufficient plastic moment of resistance to give the necessary load factor against collapse. For rolled sections the plastic moduluses are tabulated in the section tables; for a plate girder the plastic modulus may readily be calculated.

The value of the plastic moment for a continuous beam can be found from the standard methods of plastic analysis which for a uniform beam section is a very simple process. For a non-uniform section, which will lead to a minimum weight, if not necessarily minimum cost, solution a graphical method of analysis is probably the simplest way of finding the plastic moment required. Example 3.14 illustrates the design of a 3 span continuous beam.

References

1. Timoshenko, S.P. *History of Strength of Materials.* McGraw-Hill, London (1953).
2. Narayanan, R. (ed). *Beams and Beam Columns—Stability and Strength.* Applied Science Publishers, London (1982).
3. *Steelwork Design.* Guide to BS5950: Part 1:1985. Volume 1. Section Properties. Member Capacities. Constrado, London (1985).
4. Yam, L.C.P. *Design of Composite Steel–Concrete Structures.* Surrey University Press, London (1981).
5. Noble, P.W. and Leech, L.V. *Design Tables for Composite Steel and Concrete Beams for Buildings.* Constrado, London (1975).
6. Morris, L.J. and Randall, A.L. *Plastic Design.* Constrado, London (1983).

4 Axially loaded elements

4.1 Compression members

4.1.1 *Column behaviour* (1)

Members in which load is carried predominantly in compression are known variously as struts, stanchions or columns. In order to avoid confusion the word 'column' will be used here to describe compression members generally. In practice the word 'strut' is often reserved for compression members in lattice structures.

The general behaviour of axially compressed members containing residual stresses and deviations from exact straightness has been described in Section 2.2.1. Practical design of columns takes account of these theoretical aspects, modified as required for the particular characteristics of each column type. Considering first the influence of lack of straightness. The compressive strength of a pin ended column of length L with an initial bow Δ is given by the solution (lower root) to the Perry formula $(p_y - p_c)(p_E - p_c) = \eta p_E p_c$. The derivation of this formula is given in Appendix A.

The *Perry factor* η is a measure of the initial lack of straightness Δ; $\eta = \Delta(y/r^2)$, where y is the extreme fibre distance from the centroid and r is the minimum radius of gyration. Assuming that a reasonable upper limit of the ratio between Δ and L is $\Delta/L = 1/1000$ (which can be controlled by specification and verified by measurement) $\eta = (L/1000) \times (y/r^2) = \alpha\lambda$, where $\alpha = 0.001\ (y/r)$.

Table 4.1 Cross-section values y/r and α

Type of cross-section	Approximate y/r	α
Universal beams or columns		
About x–x axis	1.2	0.0012
About y–y axis	2.0	0.0020
Hollow		
Rectangular	1.3	0.0013
Circular	1.4	0.0014
Tee in plane of stem	2.8	0.0028

The parameter y/r varies significantly with the type of column cross-section (Table 4.1), leading to the conclusion that column design stresses must be related to the cross-section type.

A plot of the solution to the Perry formula for a range of values of α is shown in Figure 4.1 together with the basic yield and Euler buckling curve for $\alpha = 0.0$. Note that all curves for $\alpha > 0.0$ show an immediate reduction in p_c for $\lambda > 0.0$; there is no yield plateau. The other major influence on column strength is the presence of residual stresses which has already been described in general in Section 2.2.1. In a rolled universal column, for example, the typical residual stress pattern is shown in Figure 4.2. When the sum of the average applied compressive stress and the residual compressive stress at a point equals the

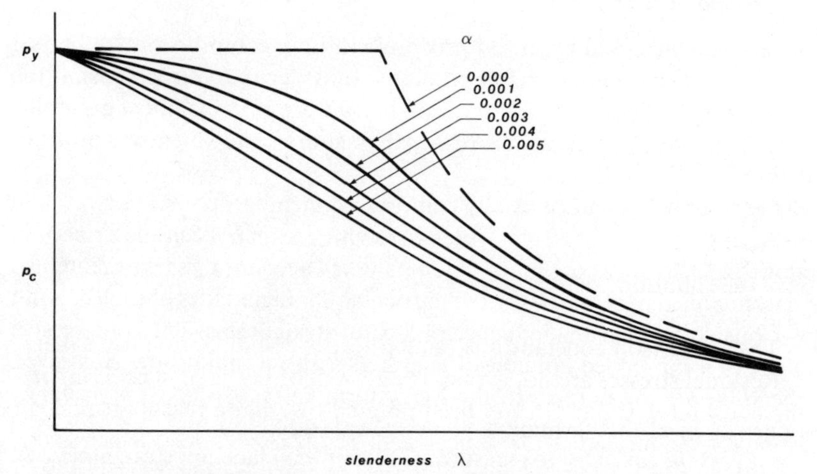

Figure 4.1 Perry formula for compressive strength

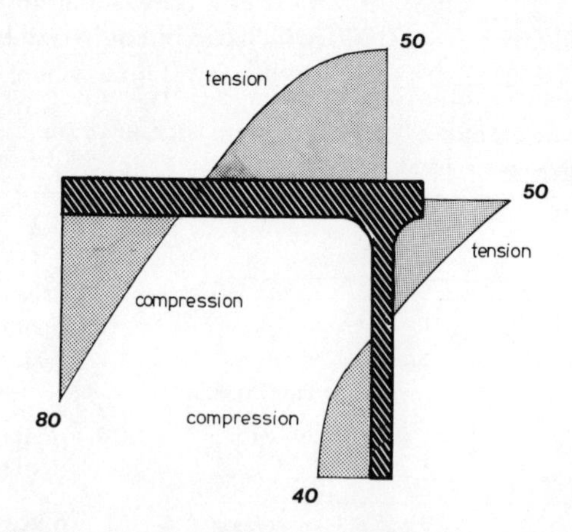

Figure 4.2 Residual stresses in a rolled section (N/mm^2)

material yield stress the column yields at that point. The yielded zones will spread inwards as the applied load is increased causing the member stiffness to reduce progressively and the column to exhibit reduced buckling stiffness. The severity of this reduction will be dependent on the pattern and magnitude of residual stresses which are a function of the shape and method of fabrication (rolling or welding) of the section.

These effects of lack of straightness and residual stress are taken account of in design by assigning compression members to one of four groups a, b, c, or d which are related to the type of cross section and method of fabrication. The column strength curve consists of two parts, an initial horizontal portion up to a slenderness λ_0 at the material design strength p_y followed by a curve of p_c calculated from the Perry equation. Thus

$$0 \leqslant \lambda \leqslant \lambda_0 \quad p_c = p_y$$

$$\lambda > \lambda_0 \quad p_c = \frac{p_E p_y}{\phi + (\phi - p_E p_y)}$$

where

$$\phi = \frac{p_y + (\eta + 1)p_E}{2}$$

λ_0, the limiting slenderness $= 0.2(\pi^2 E/p_y)^{\frac{1}{2}} = \lambda_t/5$; η, the Perry factor $= 0.001\,a(\lambda - \lambda_0) \nless$ zero.

The Robertson constant a is related to the column type (Table 4.2).

Residual stresses are not explicitly allowed for but their effect is inherent in the constant a. It has already been pointed out that a reasonable maximum value for the measure of lack of straightness is $\Delta = L/1000$ which gives $\eta = \alpha\lambda = 0.001(y/r)\lambda$. Table 4.3 gives guidance on the selection of the correct strut table.

While the derivation of the relationship between compressive strength p_c and slenderness λ by the Perry formula may be theoretically satisfactory it needs to be emphasised that the Robertson constant has been adjusted so that the resulting column curves are consistent with tests on columns of practical proportions, with realistic lack of straightness and residual stress. Figure 4.3 shows the scatter band of test results together with the column curves.

Table 4.2 Robertson constant

Type	Robertson constant a
a	2.0
b	3.5
c	5.5
d	8.0

Table 4.3 Selection of strut tables for different column types*

Column classification			Strut table	
Type		Thickness (mm)(1)	Axis of buckling	
			x–x	y–y
Rolled sections				
Hollow			a	a
I			a	b
H		up to 40 mm	b	c
		over 40 mm	c	d
I or H with welded flange cover plates		up to 40 mm	b	a
		over 40 mm	c	b
Angle Channel or tee Two laced, battened or back to back Compound			c	c
Welded sections				
Plate I or H	(2)	up to 40 mm	b	c
		over 40 mm	b	d
Box	(3)	up to 40 mm	b	d
		over 40 mm	c	c

* *Notes*

1. For thicknesses between 40 and 50 mm the value of p_c may be taken as the average of the values for thicknesses up to 40 mm and over 40 mm.
2. For welded plate I or H sections where it can be guaranteed that the edges of the flanges will only be flame cut, strut table *b* may be used for buckling about the y–y axis for flanges up to 40 mm thick, and strut table *c* for flanges over 40 mm thick.
3. 'Welded box section' includes any box section fabricated from plates or rolled sections, provided that all longitudinal welds are near the corners of the section. Box sections with welded longitudinal stiffeners are not included in this category.

The compression resistance of a column is given by $P_c = A_g p_c$, where A_g is the gross cross sectional area and p_c is the compressive strength. The compressive strength p_c depends on the slenderness λ, the design strength p_y (suitably modified if the section is slender) and the type of section. Having evaluated these factors the correct strut table can be selected and the value of p_c determined (Example 4.1).

4.1.2 *Axial load and bending*

Where bending moments exist with axial compression there is *interaction* between the two which can be expressed by an interaction equation or graphically as an interaction diagram.

Two distinct forms of failure are possible:

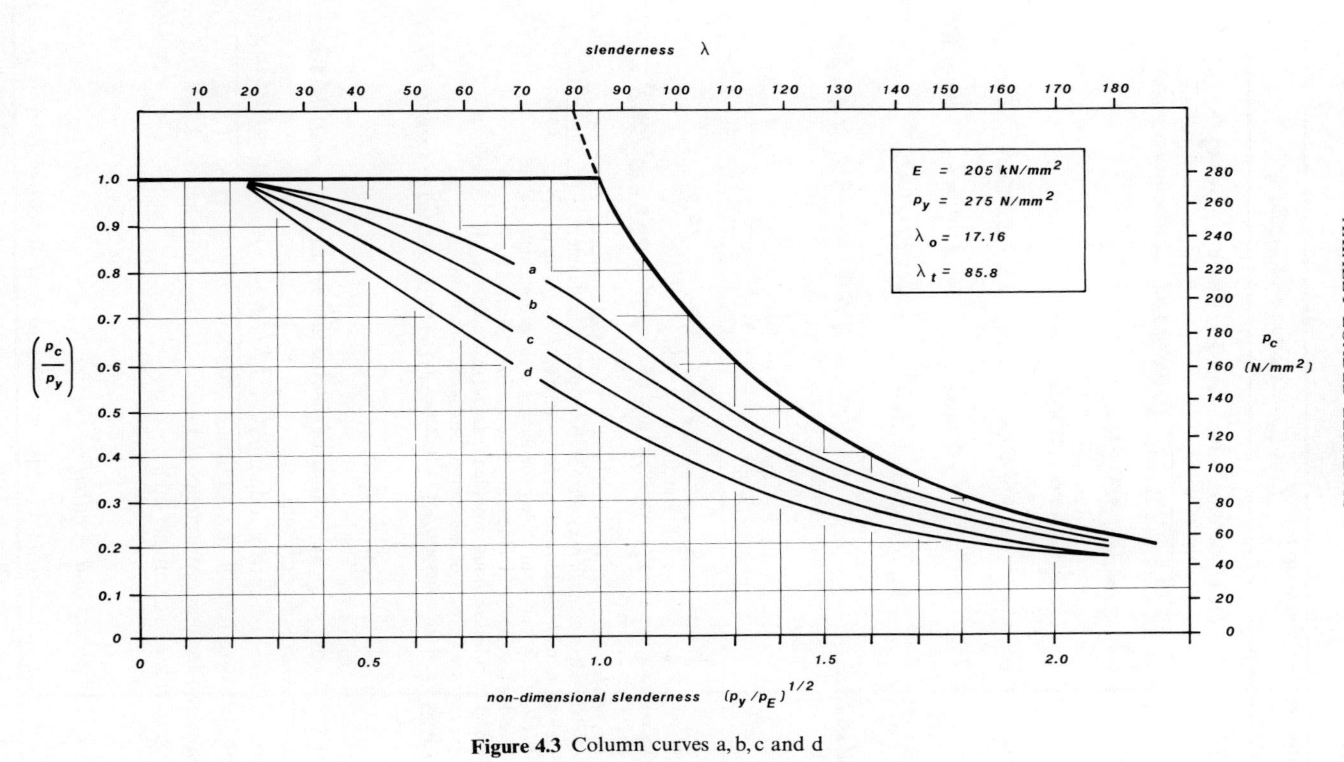

Figure 4.3 Column curves a, b, c and d

BS5950 clause	Example 4.1 Compressive strength 4.1.1
4.7.5	Determine the compressive strength of a 305 x 305 x 137 universal column in grade 50 steel.
	Effective length $L_E = 4.0$ m
	$r_x = 13.7$ cm $\quad r_y = 7.82$ cm $\quad d = 246.6$ mm
	$t = 13.8$ mm $\quad T = 21.7$ mm $\quad B = 308.7$ mm
Table 6	$p_y = 340\ N/mm^2 \quad \varepsilon = (275/340)^{1/2} = 0.899$
Table 7	$\frac{b}{T} = 7.11 \quad \frac{d}{t} = 17.9$ - not slender
	<u>x axis</u>
	$\lambda = 4.0 \times 10^2 / 13.7 = 29.2$
C.1	$p_E = \pi^2 E/\lambda^2 = \pi^2 \times 205 \times 10^3 / 29.2^2 = 2373\ N/mm^2$
C.2	$\lambda_0 = 0.2 \left(\frac{\pi^2 E}{p_y}\right)^{1/2} = 0.2 \left(\frac{\pi^2 \times 205 \times 10^3}{340}\right)^{1/2} = 15.4$
C.2	$\eta = 0.001 a (\lambda - \lambda_0)$
Table 25	rolled H section, x axis, table 27b
	$a = 3.5$
	$\eta = 0.001 \times 3.5 (29.2 - 15.4) = 0.0483$
C.1	$\phi = \frac{p_y + (\eta + 1) p_E}{2} = \frac{340 + 1.0483 \times 2373}{2} = 1414\ N/mm^2$

BS5950 clause	Example 4.1 Compressive strength 4.1.2
C.1	$p_{cx} = \dfrac{p_E p_y}{\phi + (\phi^2 - p_E p_y)^{1/2}}$
	$= \dfrac{2373 \times 340}{1414 + (1414^2 - 2373 \times 340)^{1/2}} = \boxed{322\ N/mm^2}$
	y axis
	A similar calculation, using a value of Robertson constant a = 5.5, will give:
	$p_{cy} = \boxed{262\ N/mm^2}$
	[The values in Tables 27(a) to (d) are calculated from the formulas in Appendix C of BS 5950]

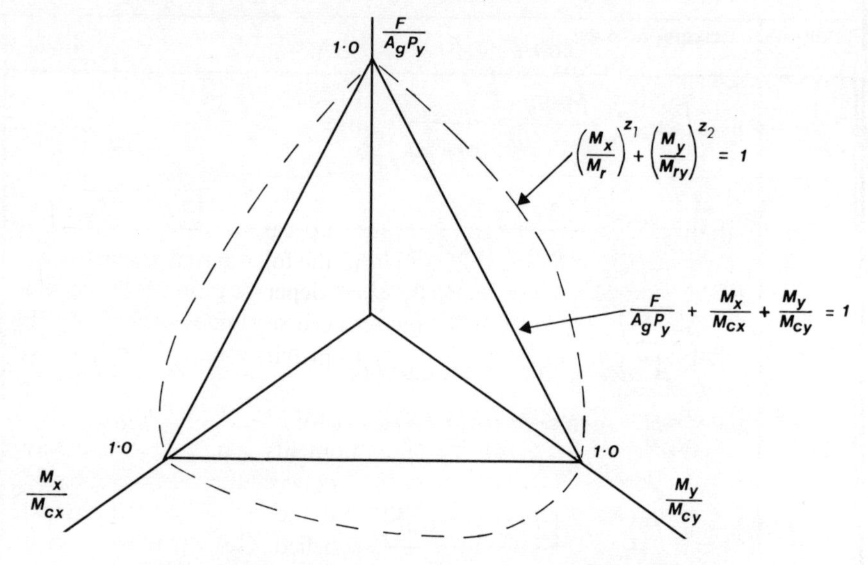

Figure 4.4 Interaction surface for local capacity

(a) a collapse at a point of high axial load and bending moment when the local capacity is exceeded
(b) an overall buckle of the member.

The analysis of either of these two forms is not simple, and is made more difficult in the case of (a) by the need to ensure that local buckling is prevented.

The interaction relationship for local capacity in a member resisting a simultaneous application of axial compression F and bending moments M_x and M_y about x and y axes respectively is shown diagrammatically in Figure 4.4. The axes show the ratios of actual loads to the member capacity in axial load or bending as relevant: *axial* $F/A_g p_y$; *moment x–x* M_x/M_{cx}; *moment y–y* M_y/M_{cy} and the member is not overloaded provided $(F/A_g p_y) + (M_x/M_{cx}) + (M_y/M_{cy}) \leqslant 1.0$. Local buckling is not critical provided that for slender members a reduced effective value of p_y is used.

Table 4.4 Reduced plastic section modulus for rolled I or H sections under axial load F

Modulus	n*	Value
S_{rx}	$\leqslant 0.2$	$(1 - 2.5n^2)S_x$
S_{rx}	> 0.2	$1.125(1 - n)S_x$
S_{ry}	$\leqslant 0.447$	$(1 - 0.5n^2)S_y$
S_{ry}	> 0.447	$1.125(1 - n^2)S_y$

*$n = F/A_g p_y$ for compression members
$n = F/A_e p_y$ for tension members

Table 4.5 Interaction constants Z_1 and Z_2

Type of section	Z_1	Z_2
I and H	2.0	1.0
Solid and closed hollow	1.67	1.67
Other	1.0	1.0

An alternative, more economical, relationship can be used for plastic or compact cross-sections only. Under axial load the full moment capacities are reduced to M_{rx} and M_{ry}, these reduced values depending on the force ratio $n = F/A_g p_y$. Tabulated expressions for the reduced section properties S_{rx} and S_{ry} are available or use can be made of the approximations in Table 4.4 for rolled I or H sections only.

The interaction relationship is $(M_x/M_{rx})^{Z_1} + (M_y/M_{ry})^{Z_2} \leqslant 1.0$. Combinations of axial load and bending moments can lead to failure by overall buckling of the member by interaction between the applied forces and moments. A simplified approach requires that the interaction relationship $(F/A_g p_c) + (mM_x/M_b) + (mM_y/p_y Z_y) \leqslant 1$ be satisfied. Note that the applied bending moments may be adjusted by the equivalent uniform moment factor m. A more exact approach is also permitted (BS5950 clause 4.8.3.3.2). For further discussion of these interaction relationships see Reference 1.

4.1.3 *Local buckling*

Premature failure of the component elements of a column can occur if limiting width to thickness ratios are exceeded. The relevant limits for a compression member at which yield will coincide with local buckling requires the section to be classified as at least semi-compact and are shown in Table 4.6. Sections not meeting these limits are classed as *slender* and their capacity may be affected by local buckling at a stress below the yield stress.

4.1.4 *Efficient design*

Because the compressive strength is a maximum for the minimum value of slenderness ratio it follows that the most efficient way to utilise the material in a column is to arrange it so as to maximise the radius of gyration. This can be

Table 4.6 Limits for semi-compact columns

Element	Limit for semi-compact element	
Outstand of flange	Rolled	$b/T \leqslant 15\varepsilon$
	Welded	$b/T \leqslant 13\varepsilon$
Web	Rolled	$d/T \leqslant 39\varepsilon$
	Welded	$d/T \leqslant 28\varepsilon$

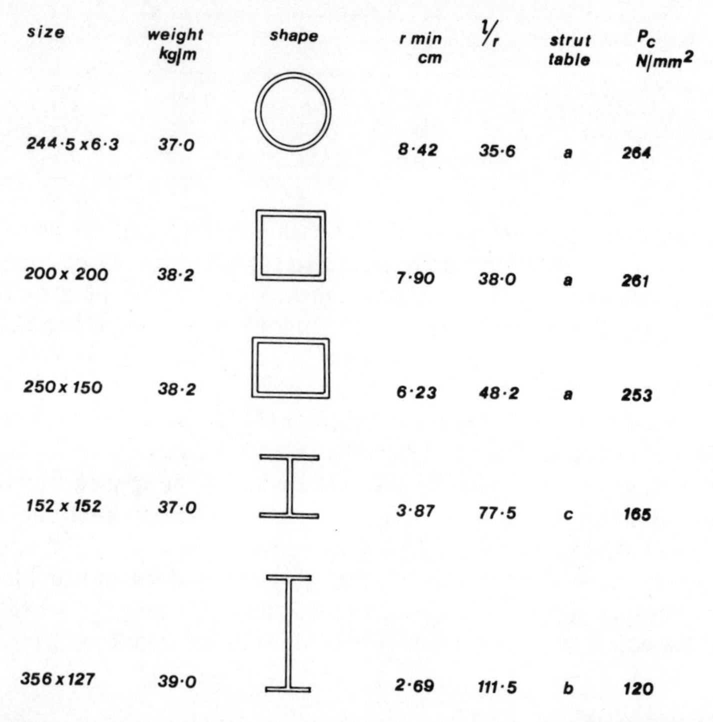

Figure 4.5 Comparison of various section shapes as compression members (Slenderness L_E/r based on effective length of 3.0 m. Compressive strength from strut tables, BS5950. for P_y of 275 N/mm^2)

achieved by disposing the material as far as possible from its centroid. For this reason a thin-walled hollow section is a very efficient column, subject to the proviso that if the wall is too thin it will buckle locally.

Theoretically, a thin-walled circular section is the optimum shape but there will be circumstances in which a rectangular hollow section will be more efficient than a circular section, as, for example, when bending is combined with compression. Figure 4.5 shows a comparison of five different rolled sections of approximately equal cross-sectional area listed in descending order of minimum radius of gyration from which it will be observed that the circular hollow section has a minimum radius of gyration over three times and a compressive strength over twice that of the universal beam.

4.1.5 *Column design*

The compressive strength p_c is determined by the slenderness ratio $\lambda =$ (effective length/minimum radius of gyration) but the minimum radius of gyration cannot be calculated until the cross-section has been fixed. The

Table 4.7 Effective lengths for various end restraint conditions

Conditions of restraint at the ends (in plane under consideration)		Effective length factor
Effectively held in position at both ends	Restrained in direction at both ends	0.7
	Partially restrained in direction at both ends	0.85
	Restrained in direction at one end	0.85
	Not restrained in direction at either end	1.0

One end	Other end		
	Position	*Direction*	
Effectively held in position and restrained in direction	Not held	Effectively restrained	1.2
		Partially restrained	1.5
		Not restrained	2.0

One end		Other end		Effective length factor
Position*	Direction†	Position	Direction	
H	R	H	R	0.7
H	PR	H	PR	0.85
H	R	H	N	0.85
H	N	H	N	1.0
H	R	N	R	1.2
H	R	N	PR	1.5
H	R	N	N	2.0

* Position: H, held; N, not held.
† Direction: R, restrained; PR, partially restrained; N, not restrained.

familiar circular argument of design is apparent here but the circle can be cut by making a guess and then, if needful, improving on it. As will be seen later, tables from which direct design is possible are available for the standard rolled sections.

Estimation of the slenderness ratio of a column of known radius of gyration depends on the ability to assign an effective length to it. If the actual length is L (measured from the centre of supports or supporting members) then the effective length $L_E = kL$, where k, the effective length factor, depends on the positional and directional restraint given to the ends of the column. Some values of effective length factor are shown in Table 4.7. Relating these theoretical end conditions to those actually found in steel structures is not always easy. Useful guidance however is given in British Standard 5950.

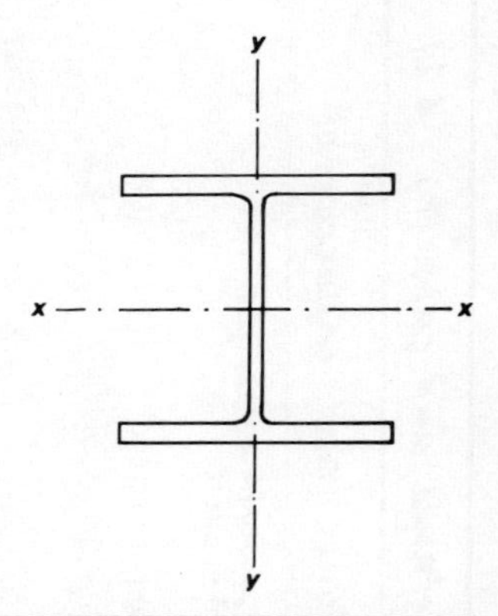

Resistances and capacities for steel grade 50

Designation and Capacities		Compression resistance P_{cx}, P_{cy} (kN) and buckling resistance moment M_b (kNm) for effective length L_E (m) Reduced moment capacity M_{rx}, M_{ry} (kNm) for ratios of axial load to axial load capacity (F/P_z)												
	L_E(m)	1.5	2.0	2.5	3.0	3.5	4.0	5.0	6.0	7.0	8.0	9.0	10.00	11.0
	P_z	0.05	0.10	0.15	0.20	0.25	0.30	0.35	0.40	0.45	0.50	0.55	0.60	0.65
305 × 305 × 137	P_{cx}	5820	5950	5890	5810	5720	5640	5450	5230	4970	4670	4330	3970	3590
$P_z = 5950$	P_{cy}	5820	5600	5370	5130	4870	4590	3980	3370	2810	2340	1960	1650	1410
$M_{cx} = 782$	M_b	782	782	782	782	759	732	681	633	589	549	512	479	450
$M_{cy} = 282$	M_{bs}	782	782	782	782	782	782	775	732	687	640	591	542	494
$p_y.Z_y = 235$	M_{rx}	777	763	740	707	667	626	584	541	499	455	412	368	323
	M_{ry}	282	282	282	282	282	282	282	282	282	282	282	269	245

Note: F = factored axial load
M_b is obtained using an equivalent slenderness = n.u.v.L_E/r with $n = 1.0$.
M_{bs} is obtained using an equivalent slenderness = $0.5L/r$
Values have not been given for P_{cx} and P_{cy} if the values of slenderness are greater than 180.

Figure 4.6 Resistances and capacities of a universal column. (source: Steel Construction Institute)

BS5950 clause	Example 4.2 Column resistance and capacity 4.2.1
	Determine the resistance and capacity of the universal column of example 4.1
	Classification
	No reduction in design strength as the
Table 7	section is not slender ($d/t < 39 \times 0.899$)
	Strut table
Table 25	$x-x$ 27b $y-y$ 27c
	x axis
Table 27b	$p_{cx} = 322\ N/mm^2$
	$P_{cx} = 322 \times 175 \times 10^2 \times 10^{-3}$ = **5640 kN**
	($A = 175\ cm^2$)
	y axis
	$p_{cy} = 262\ N/mm^2$
	$P_{cy} = 262 \times 175 \times 10^{-1}$ = **4590 kN**
	Buckling resistance moment
	$\lambda = L_E/r_y = 51.2$
	From properties table
	$u = 0.851$, $x = 14.1$
	$\lambda/x = 51.2/14.1 = 3.63$, $N = 0.5$
Table 14	$v = 0.88$, $n = 1.0$
	$\lambda_{LT} = 1.0 \times 0.851 \times 0.88 \times 51.2 = 38.3$
Table 11	$p_b = 318.1\ N/mm^2$
	$M_b = 318.1 \times 2300 \times 10^{-3}$ = **732 kNm**

BS5950 clause	Example 4.2 Column resistance and capacity 4.2.2
	Squash load
	$P_z = 175 \times 340 \times 10^{-1} = $ **5950 kN**
4.8	Axial load and bending
	$S_{xx} = 2300\,cm^3$, $S_{yy} = 1050\,cm^3$, $Z_{yy} = 691\,cm^3$
	Moment capacities
	$b/T = 7{\cdot}11 < 8{\cdot}5 \times 0{\cdot}899 = 7{\cdot}64$
Table 7	plastic section
4.2.5	$M_{cx} = 340 \times 2300 \times 10^{-3} = $ **782 kNm**
	$M_{cy} = 1{\cdot}2 \times 340 \times 691 \times 10^{-3} = $ **282 kNm**
	$(340 \times 1050 \times 10^{-3} = 357\,kNm)$
	Check reduced capacities if ratio
	axial load/squash load, $F/P_z = 0{\cdot}3 = n$
	$S_{rx} = 248(1-0{\cdot}3)(10{\cdot}3+0{\cdot}3)$ $[n > 0{\cdot}219]$
	$= 248 \times 0{\cdot}7 \times 10{\cdot}6 = 1840\,cm^3$
	$S_{ry} = 1700(1-0{\cdot}3)(0{\cdot}563+0{\cdot}3)$ $[n > 0{\cdot}254]$
	$= 1700 \times 0{\cdot}7 \times 0{\cdot}863 = 1030\,cm^3$
	(formulas from section tables)
4.8.2	$M_{rx} = 340 \times 1840 \times 10^{-3} = $ **626 kNm**
	M_{ry} lesser of
	$340 \times 1030 \times 10^{-3} = 350\,kNm$
	or **282 kNm**

4.1.6 *Member capacity tables*

For the commonly available rolled sections tables of compression resistance are available from which direct design is possible, not only for axially loaded members but also for members carrying a combination of axial load and bending moment (2). Extracts are shown in Figure 4.6. Example 4.2 illustrates the way in which these tabular entries are derived.

4.1.7 *Columns in building frames*

Load is transferred from floor and roof beams into the columns through beam to column connections, some examples of which are illustrated in Figure 4.7. It will be apparent that all these connections must impart some eccentricity of

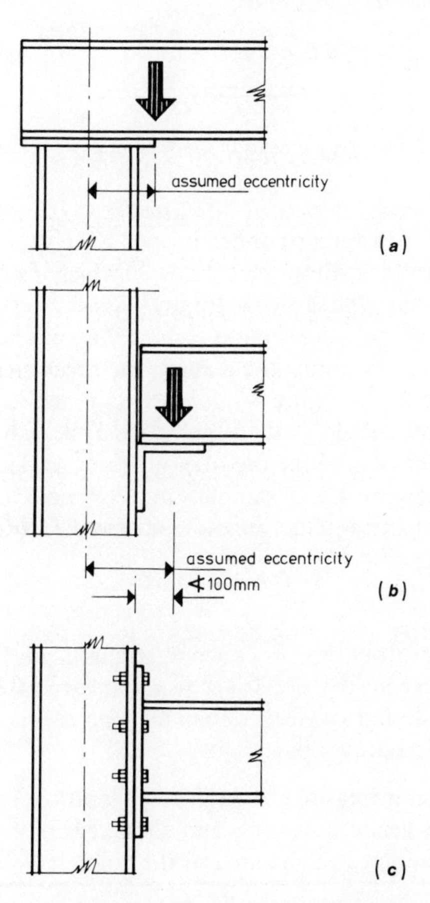

Figure 4.7 Eccentric connections (*a*) Cap connection. Reaction assumed to be at the edge of packing if used, otherwise at face of column; (*b*) Simple end cleat. Reaction assumed to be at the greater of 100 mm from column face or centre of bearing; (*c*) Rigid connection

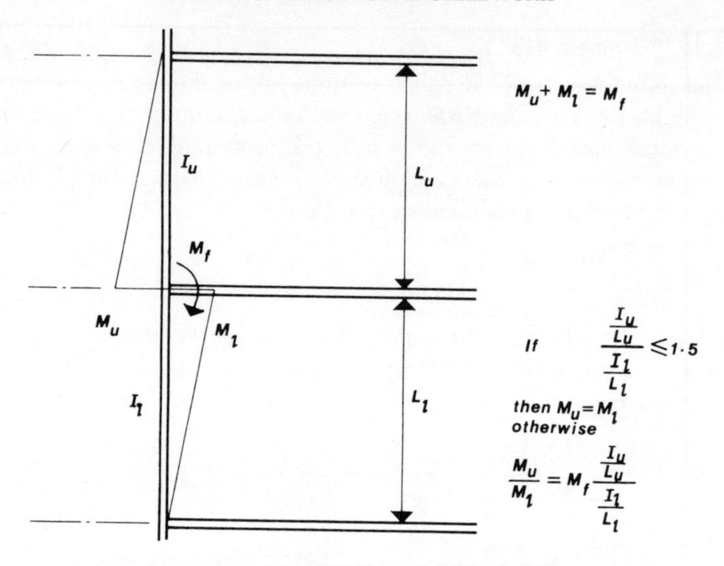

Figure 4.8 Bending moments in simple multi-storey construction

load of the column; even the cap connection will not give concentric loading to the column because of the rotation of the beam which it supports. In addition rigid connections of the type shown in Figure 4.7c will, by frame action, induce bending into the column. For simple connections which do no more than transfer the beam reaction into the column the eccentricity of load may be calculated by reference to Figure 4.7a or b. For jointed continuous columns the bending moment caused by the eccentricity may be distributed between the column lengths above and below the junction under consideration in accordance with Figure 4.8 (Example 4.3). For rigid connections elastic analysis will be required to determine the distribution of bending moments.

4.1.8 *Cased columns*

Subject to certain restrictions on concrete strength, steel section, minimum cover and reinforcement (Figure 4.9) it is possible to take advantage of the increased strength given to a steel column by concrete encasement. The extra strength is composed of two parts:

(a) The concrete increases the radius of gyration of the combination reducing the slenderness ratio and so increasing the design strength.
(b) The concrete increases the area of the column.

The exact manner in which the allowable load on a concrete-cased column is calculated is to some extent based on empirical formulae. In British Standard 5950 the following apply:

BS5950 clause	Example 4.3 Multi-storey columns 4.3.1

350 kN

2nd floor

3·0m — upper

1st floor

self weight 30 kN

5·0m — lower

110 kN

20 kN

Loads are factored

Try a 203×203×52 UC - grade 43

Eccentricity of beam loads

4.7.6(a)(3)

$e_1 = 100 + \frac{206.2}{2} = 203\text{mm}$

$e_2 = 100 + \frac{8}{2} = 104\text{ mm}$

e_1, e_2, 8, 206·2

4.7.7

Divide moments

upper stiffness ∝ 1/3, lower stiffness ∝ 1/5

$\frac{1/3}{1/5} > 1.5$

∴ divide moment at 1st floor level

above $M_u = M\left(\frac{1/3}{(1/3 + 1/5)}\right) = 0.62\,M$

below $M_L = M\left(\frac{1/5}{(1/3 + 1/5)}\right) = 0.38\,M$

BS5950 clause	Example 4.3 Multi storey columns 4.3.2
	Consider loading in lower column length just below 1st floor:
	total factored axial load
	$F = 350 + 2(110 + 20) + 30 = \boxed{640\ kN}$
	$M_x = 110 \times 0.203 \times 0.38 = \boxed{8.49\ kN\ m}$
	$M_y = 20 \times 0.104 \times 0.38 = \boxed{0.790\ kN\ m}$
Table 6	$p_y = 275\ N/mm^2 \quad (T < 16)$
	P_z (squash load) $= 275 \times 66.4 \times 10^{-1} = \boxed{1830\ kN}$
	$F/P_z = 640/1830 = 0.35$
	Section values from Constrado Guide vol 1, p176, for an effective length of 5.0m
	P_{cy} 865 kN $\quad p_y Z_y$ 47 kNm
	M_b 117 kNm
	M_{rx} 116 kN m
	M_{ry} 57 kNm
4.8.3	Check local capacity.
.2(b)	plastic section
	$\left(\frac{M_x}{M_{rx}}\right)^{z_1} = \left(\frac{8.49}{116}\right)^2 = 0.0054$
	$\left(\frac{M_y}{M_{ry}}\right)^{z_2} = \left(\frac{0.79}{57}\right) = 0.0139$
	$0.0192 < 1.0$

BS5950 clause	Example 4.3 Multi-storey columns 4.3.3
4.8.3.3	Check overall buckling
4.8.3.3.1	$\frac{F}{A_g p_c} = \frac{640}{865} = 0.740$
	$\frac{m M_x}{M_b} = \frac{8.49}{117} = 0.073$ (see notes 1-2)
	$\frac{m M_y}{p_y Z_y} = \frac{0.79}{47} = 0.017$ (see note 1)
	$0.830 < 1.0$
	Note 1. m taken, conservatively, as 1.0
	Note 2. M_{bs} could be substituted for M_b - see clause 4.7.7

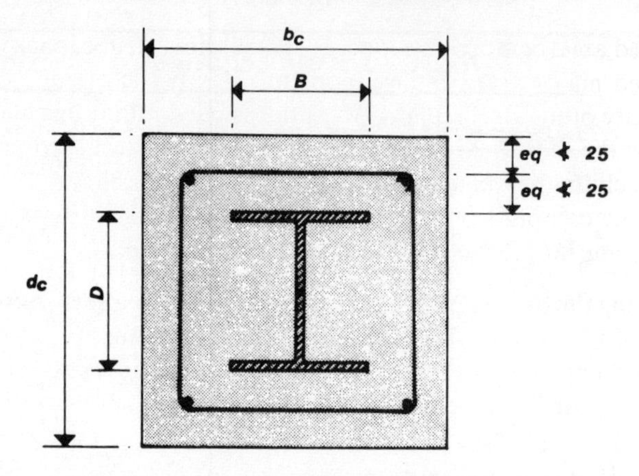

Figure 4.9 Concrete cased I section. *Concrete*: ordinary, dense, structural, at least grade 20 *Reinforcement*: not less than 5 mm diameter in the form of a cage of longitudinal bars held by closed links spaced not more than 200 mm apart. *Cover*: 50 mm minimum to outer faces and edges of steel section

The compression resistance P_c is composed of:

steel	$A_g p_c$
concrete	$0.45 A_c f_{cu} p_c / p_y$

but must not exceed the short strut (squash) capacity P_{cs} given by the sum of the components:

steel	$A_g p_y$
concrete	$0.25 A_c f_{cu}$

Certain restrictions apply to the size and disposition of steel sections which can be used in a cased section. The details which follow apply to cased I or H sections, welded or fabricated, with equal flanges.

(a) The effective length, L_E, of the cased section is the least of: $40b_c$, $100b_c^2/d_c$ or $250r$, where b_c and d_c are defined in Figure 4.9 and r is the minimum radius of gyration of the steel section alone.
(b) The radius of gyration r_y about the axis in the plane of the web is: $0.2b_c$ but not more than $0.2(b + 150)$ mm. If r_y for the steel section alone is greater than that for the composite section the steel section value may be used.
(c) The radius of gyration r_x about the axis parallel to the planes of the flanges is that of the steel section alone.

When a cased column has to resist combined axial compression and bending moment it must satisfy the following relationships:

(a) for capacity, $(F_c/P_{cs}) + (M_x/M_{cx}) + (M_y/M_{cy}) \leqslant 1.0$, where F_c is the applied axial compressive load, P_{cs} is the short strut capacity, M_x is the applied major axis bending moment, M_{cx}, is the major axis moment capacity of the steel section, M_y is the applied minor axis moment and M_{cy} is the minor axis moment capacity of the steel section

(b) for buckling resistance, $(F_c/P_c) + (mM_x/M_b) + (mM_y/M_{cy}) \leqslant 1.0$, where P_c is the compression resistance, m is the equivalent uniform moment factor and M_b is the buckling resistance moment.

For calculating the buckling resistance moment of a cased section the radius of gyration r_y may be taken as the greater of $0.2(B + 100)$ mm or r_y of the uncased section. All other properties are to be those of the uncased section. The value of M_b thus computed should not exceed 1.5 M_b for the uncased section (Example 4.4).

4.1.9 *Concrete-filled columns*

By filling a steel hollow section with concrete a substantial increase in load-carrying capacity can be achieved without any increase in the physical size of the hollow section. In addition the amount of fire protection required for the filled section will be less than that for the unfilled one. Local buckling in the wall of the steel section will be restrained by the concrete infill.

Design of these filled columns is experimental and computer based; safe load tables are available (3).

4.1.10 *Compound columns*

A combination of rolled sections forming a built-up or compound column has advantages for larger loads; some practical combinations are illustrated in Figure 4.10. The interconnection between the component sections *can* be

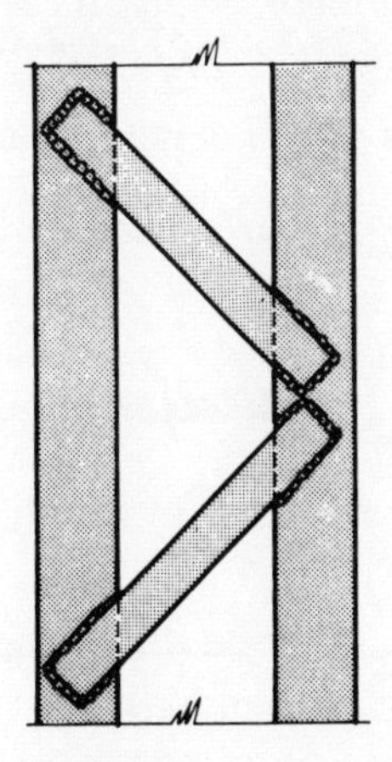

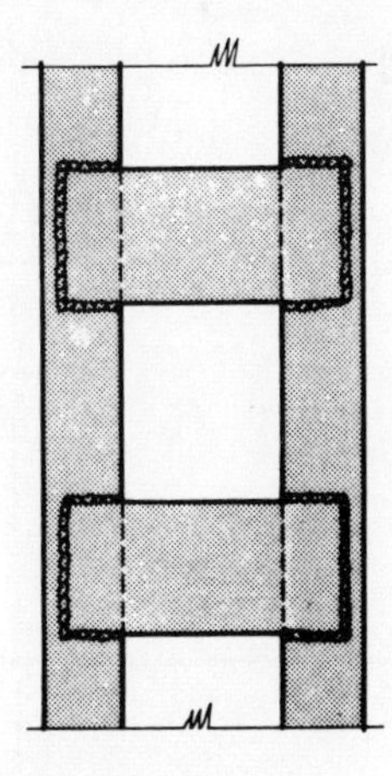

Figure 4.10 Laced (left) and battened (right) columns

BS5950 clause	Example 4.4 Cased strut 4.4.1
	<u>305 × 305 × 283 UC</u>, grade 43 steel
	effective length 5·0 m
Table 6	$T = 44{\cdot}1$ mm , $p_y = 245$ N/mm^2
	<u>Uncased capacity</u>:
	From Constrado Guide, vol 1, p 86
	$P_{cx} = 8030$ kN ; $P_{cy} = 5940$ kN
4.14	<u>Cased</u>
4.14.1	$B = 321{\cdot}8$ mm
	$b_c = 430$ mm
	concrete $f_{cu} = 25$ N/mm^2
	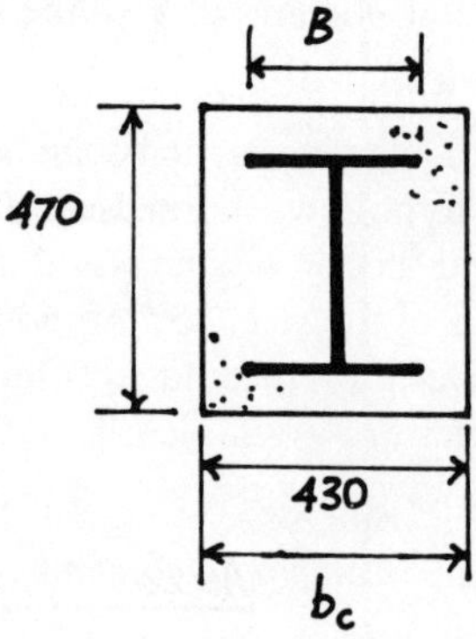
4.14.3(a)	<u>r_y</u>:
	lesser of
	$0{\cdot}2\,(B + 150) = 0{\cdot}2\,(321{\cdot}8 + 150) = 9{\cdot}44$ cm
	$0{\cdot}2\,b_c = 0{\cdot}2 \times 430 = 8{\cdot}60$ cm
	$\therefore r_y = \boxed{8{\cdot}60 \text{ cm}}$ (steel section $r_y = 8{\cdot}25$ cm)
	r_x:
	$r_x = \boxed{14{\cdot}8 \text{ cm}}$ (steel section)

BS5950 clause	Example 4.4 Cased strut 4.4.2
4.14.3(b)	Compression resistance
	a. Short strut capacity
	concrete $0.25 \times 25 \times (470 \times 430) \times 10^{-3} = 1260$ kN
	steel $360 \times 245 \times 10^{-1} = 8820$ kN
	$P_{cs} = 10{,}100$ kN
	b. Compression resistance
	$\lambda = 5 \times 10^3 / 8.6 \times 10 = 58.1$
Table 25 Table 27d	$p_c = 170$ N/mm^2
	concrete $0.45 \times 25 \times \frac{170}{245} (470 \times 430) \times 10^{-3} = 1580$ kN
	steel $360 \times 170 \times 10^{-1} = 6120$ kN
	$P_c = 7700$ kN
4.14	Axial load and moment
	An eccentric load of P kN acts at 100 mm from the flange of the steel section, on the y-y axis
	$F_c = P$ kN
	$M_x = P(100 + 182.7) \times 10^{-3} = 0.283$ kN m

BS5950 clause	Example 4.4 Cased strut 4.4.3
4.14.4(a)	For capacity
	M_{cx} = 1250 kNm (Constrado Guide)
	$\frac{F_c}{P_{cs}} = \frac{P}{10100} = 99\times10^{-6}\,P$
	$\frac{M_x}{M_{cx}} = \frac{0.283P}{1250} = 226\times10^{-6}\,P$
	$325\times10^{-6}\,P$
	$325\times10^{-6}\,P = 1.0$, P = 3080 kN
4.14.4(b)	Buckling resistance
	M_b for uncased section, $n = 1.0$
	M_b = 1250 kNm (Constrado Guide)
	$\frac{F_c}{P_c} = \frac{P}{7700} = 130\times10^{-6}\,P$
	$\frac{M_x}{M_b} = \frac{0.283P}{1250} = 226\times10^{-6}\,P$
	$356\times10^{-6}\,P$
	$356\times10^{-6}\,P = 1.0$, P = 2810 kN
	Buckling resistance determines value of P as 2810 kN

BS5950 clause	Example 4.5 Compound struts 4.5.1
	305 × 102 × 46.18 channels
	$C_y = 2.66\,cm$ $A = 58.8\ cm^2$
	$I_{xx} = 8210\,cm^4$ $I_{yy} = 499\,cm^4$
	$r_{xx} = 11.8\,cm$ $r_{yy} = 2.91\,cm$
	$a = 10.16 - 2.66 = 7.5\,cm$
4.7.13.1	a. **Back-to-back, separated**
	$s = 15\,mm$
	$r_{xx} = 11.8\,cm$
	$I_{yy} = 2 \times 499 + 2 \times 58.8 \times 3.41^2 = 2370\ cm^4$
	$r_{yy} = (2370/2 \times 58.8)^{1/2} = \boxed{4.49\,cm}$
	b. **Toe to toe**
	$r_{xx} = 11.8\,cm$
	$I_{yy} = 2 \times 499 + 2 \times 58.8 \times 7.5^2 = 7610\ cm^4$
	$r_{yy} = (7610/2 \times 58.8)^{1/2} = \boxed{8.04\,cm}$

BS5950 clause	Example 4.5 Compound struts	4.5.2

c. Laced or battened

i Toes in

spacing between toes a cm

$r_{xx} = 11{\cdot}8$ cm

$I_{yy} = 2 \times 499 + 2 \times 58{\cdot}8 \left(\frac{a}{2} + 7{\cdot}5\right)^2$

$r_{yy} = \left(\frac{I_{yy}}{2 \times 58{\cdot}8}\right)^{1/2}$

a cm	r_{yy} cm
0	8·04
10	12·8
20	17·7

spacing to give $r_{xx} = r_{yy} = 11{\cdot}8$ cm:

$a = 7{\cdot}87$ cm

ii Toes out

$r_{xx} = 11{\cdot}8$ cm

$I_{yy} = 2 \times 499 + 2 \times 58{\cdot}8 \left(\frac{a}{2} + 2{\cdot}66\right)^2$

$r_{yy} = \left(\frac{I_{yy}}{2 \times 58{\cdot}8}\right)^{1/2}$

spacing to give $r_{xx} = r_{yy} = 11{\cdot}8$ cm

$a = 17{\cdot}6$ cm

BS5950 clause	Example 4.6 Laced column 4.6.1
4.7.8	305 × 102 × 46·18 channels spaced 80 mm between toes Effective length, L_E, of laced column about x-x or y-y axis 7·0 m, grade 43 steel
	a. From example 4·5: $r_{min} = r_{xx} = 11{\cdot}8$ cm $\lambda = \frac{L_E}{r} = \frac{7\times10^3}{11{\cdot}8\times10} = 59{\cdot}3$
Table 6 Table 25 Table 27(c)	$p_y = 275$ N/mm² (T<16 mm) $p_c = 202$ N/mm² capacity of laced column $2\times58{\cdot}8\times202\times10^{-1}$ = **2380 kN**
	b. Check portion of single channel between intersections of lacings: $c = 80 + 2\times75 = 230$ mm $b = 2(230\tan\alpha) = 460\tan\alpha$ If b is the effective length, slenderness ratio = $460\tan\alpha/2{\cdot}91\times10 = 15{\cdot}8\tan\alpha$
4.7.8(9)	Slenderness ratio must not exceed :- i. 50 ii. slenderness ratio of column / 1·4

BS5950 clause	Example 4.6 *Laced column* 4.6.2
	$59.3/1.4 = 42.4$
	$\alpha = \tan^{-1}(42.4/15.8) = 69.6^\circ$
4.7.8(e)	Angle of bars, α, must be between 40° and 70° - adopt $\alpha = 45^\circ$, giving
4.7.8(h)	slenderness ratio of $15.8 < 180$
	c. <u>Design lacing bars</u>
4.7.8(h)	length of bar between inner end welds
	$= 80/\cos 45^\circ = 113$ mm
4.7.8(i)	transverse shear $= 0.01 \times 2380 = 23.8$ kN
	axial force in bar $= 23.8/2\times\cos 45^\circ = 16.8$ kN
	try flats 25x6
	$r = 6/(12)^{1/2} = 1.73$ mm
	$\lambda = 113/1.73 = 65.3$
Table 25 Table 27(b)	$p_c = 211$ N/mm^2
4.7.4	compression resistance of 1 flat
	$= 25 \times 6 \times 211 \times 10^{-3} = \boxed{31.7 \text{ kN}} > 16.8$ kN

BS5950 clause	Example 4.7 Discontinuous angle struts 4.7.1
4.7.10	<u>100×75×12 unequal angle</u>
	A 19.7 cm^2 r_{xx} 3.1 cm
	I_{xx} 189 cm^4 r_{yy} 2.14 cm
	I_{yy} 90.2 cm^4 r_{vv} 1.59 cm
	I_{vv} 49.5 cm^4
	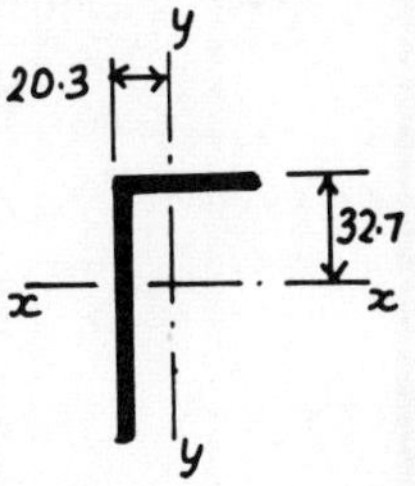
4.7.10.1	Assume length of strut between intersections is 2.0 m, long leg attached.
	<u>Classification</u>
Table 6	Grade 43, $t < 16$ mm, $p_y = 275$ N/mm^2
	$\varepsilon = 1.0$
Table 7	$b/T = 100/12 = 8.33 < 8.5\varepsilon$
	section is not slender.
4.7.10.2	a. <u>Single angle, single bolted</u>
	slenderness greater of:
	$1.0\,L/r_{vv} = 2.0\times10^3/1.59 = 126$
	$(0.7\,L/r_{aa}) + 30 = (0.7\times2.0\times10^3/2.14)+30 = 95.4$
	$\lambda = 126$
Table 27(c)	$p_c = 90$ N/mm^2
4.7.10.2	compression resistance P_c
	$= 0.8\times19.7\times90\times10^{-1} =$ **142 kN**

BS5950 clause	Example 4.7 *Discontinuous angle struts* 4.7.2
4.7.10.2(a)	b. *single angle, two bolts in line*
	slenderness greater of:
	$0.85 \times 2.0 \times 10^2 / 1.59 = 107$
	$(0.70 \times 2.0 \times 10^2 / 2.14) + 30 = 95.4$
	$\lambda = 107$
Table 27(c)	$p_c = 115\ N/mm^2$
	$P_c = 19.7 \times 115 \times 10^{-1} = \boxed{227\ kN}$
4.7.10.3(b)	c *double angle, two bolts in line, connected to both sides of a gusset*
	$I_{xx} = 2 \times 189 = 378\ cm^4$
	$I_{yy} = 2 \times 90.2 + 2(19.7 \times 3.03^2) = 542 cm^4$
	$r_{xx} = (378 / 2 \times 19.7)^{1/2} = 3.10\ cm$
	$\lambda = 0.85 \times 2 \times 10^2 / 3.1 = 54.8$
	or $(0.7 \times 2 \times 10^2 / 3.1) + 30 = 75.2$
	$\lambda = 75.2$
Table 27c	$p_c = 171\ N/mm^2$
	$P_c = 2 \times 19.7 \times 171 \times 10^{-1} = \boxed{674\ kN}$

continuous but there is no necessity to do more than provide sufficient intermittent connection to avoid local instability of a component between these connections. The components may be placed in contact or separated by a small distance or may be widely separated. Design rules for compound columns are formulated so as to provide adequate strength in the connections joining the components and to prevent local instability.

The increased load capacity of a compound column is demonstrated in Example 4.5. By varying the separation of the component channels it is possible to provide equal radii of gyration about both axes. This will be the most efficient arrangement if the effective length for either axis is the same.

The design rules for laced, battened and back-to-back compression members are rather complex. The reader is referred to BS 5950 for full details (Clauses 4.7.8, 4.7.9 and 4.7.10); Example 4.6 illustrates the use of these clauses.

4.1.11 *Angle struts*

In lightweight lattice girders economical compression members (struts) can be fabricated from angles either singly or in pairs. Because these struts are generally not loaded along their centroidal axes (their lack of symmetry makes a centroidal connection difficult) some eccentricity of axial loading is often inevitable.

Where angle struts are not continuous between connections there are design rules in BS5950 which avoid the necessity of taking account of eccentricity of loading.

Tables of compression resistance for all four categories of angle strut have been published (2). Methods of calculation are illustrated in Example 4.7.

4.2 Tension members

Members axially loaded in tension are found principally in lattice frames, in which they are often called ties. In contrast to compression members the disposition of the material in a tie has no effect on its structural efficiency so that compact sections such as rods may be used without reduction in allowable stress.

The general loading case for a tie, combining tension and bending, is dealt with by interaction relationships similar to those for columns.

Bending stress caused by eccentricity of loading will occur in sections which are not loaded along their centroidal axes.

Simple tension members composed of angles, channels or T-sections can be designed by taking account of the eccentricity of loading by the use of an empirical effective area in much the same way as for struts. Example 4.8 illustrates these points.

BS5950 clause	Example 4.8 Angle ties	4.8.1
4.6	100×75×12 unequal angle	
Table 6	grade 43 steel, $p_y = 275\ N/mm^2$	
4.6.3.2 NOTE	areas: long leg $12(100-6) = 11{\cdot}3\ cm^2$	
	short leg $12(75-6) = 8{\cdot}28\ cm^2$	
	a. Single angle connected through one leg	
	deduct 1×22 mm diameter hole in connected leg, $12 \times 22 \times 10^{-2} = 2{\cdot}64\ cm^2$	
	i short leg connected	
	$a_1 = 8{\cdot}28 - 2{\cdot}64 = 5{\cdot}64\ cm^2$	
	$a_2 = 11{\cdot}28\ cm^2$	
4.6.3.1	$3a_1/(3a_1 + a_2) = 3\times5{\cdot}64/(3\times5{\cdot}64 + 11{\cdot}28) = 0{\cdot}600$	
	effective area	
	$5{\cdot}64 + 0{\cdot}600 \times 11{\cdot}28 = 12{\cdot}4\ cm^2$	
	$P_t = 12{\cdot}4 \times 275 \times 10^{-1} =$ **341 kN**	
	ii long leg connected	
	$a_1 = 11{\cdot}28 - 2{\cdot}64 = 8{\cdot}64\ cm^2$	
	$a_2 = 8{\cdot}28\ cm^2$	
	$3a_1/(3a_1 + a_2) = 3\times8{\cdot}64/(3\times8{\cdot}64 + 8{\cdot}28) = 0{\cdot}758$	
	effective area	
	$8{\cdot}64 + 0{\cdot}758 \times 8{\cdot}28 = 14{\cdot}9\ cm^2$	
	$P_t = 14{\cdot}9 \times 275 \times 10^{-1} =$ **409 kN**	

BS5950 clause	Example 4.8 Angle ties 4.8.2
	b. a pair of angles connected to the same
4.6.3.2	side of a gusset by welding.
	long legs connected
	connected leg area
	$2 \times 11.3 \quad = \quad 22.6\,cm^2$
	unconnected leg area
	$2 \times 8.28\,\{5 \times 22.6/(5 \times 22.6 + 2 \times 8.28)\} \quad = \quad 14.4\,cm^2$
	total area = $22.6 + 14.4 \quad = 37.0\,cm^2$
	$P_t = 37 \times 275 \times 10^{-1} \quad = \boxed{1020\ kN}$
	c. as for b but connected by a single
	bolt in each short leg, 22 mm dia. hole:
	connected leg area
	$22.6 - 2 \times 2.64 \quad = \quad 17.3\,mm^2$
	unconnected leg area
	$2 \times 8.28\,\{5 \times 17.3/(5 \times 17.3 + 2 \times 8.28)\} \quad = \quad 13.9\,cm^2$
	total area = $17.3 + 13.9 \quad = 31.2\,cm^2$
	$P_t = 31.2 \times 275 \times 10^{-1} \quad = \boxed{858\ kN}$

References

1. Narayanan, R. (ed.) *Axially Compressed Structures–Stability and Strength.* Applied Science Publishers, London (1982).
2. *Steelwork Design.* Guide to BS5950: Part 1:1985. Volume 1. Section Properties. Member Capacities. Constrado, London (1985).
3. *Construction with Hollow Steel Sections.* British Steel Corporation, Tubes Division, Corby (1984).

5 Steelwork connections

5.1 General

The fact that steel members can be readily connected is at one and the same time both a useful and a potentially embarrassing characteristic of the structural material. It is useful in that large, complex structures of monumental size can be fabricated from more manageable components, with the advantages in transportation and construction which this involves, but embarrassing in that the connections may, if not designed with care, be a source of weakness in the finished structure, not only in their structural action but also because they may be the focus of corrosion and aesthetically unpleasing. Although the importance of good connections cannot be over-emphasised, it is true to say that often they are not given the attention they merit, sometimes being considered a subsidiary topic to be dealt with after the structural elements have been designed. Partly this is due to the fact that, whereas the design of main members has reached an advanced stage, based on theories which have been developed and refined, the behaviour of connections is often so complex that theoretical considerations are of little use in practical design. It is considered essential here to start from basic principles, and to show how the complex actions in connections are simplified into more convenient design methods. The designer will then be able to keep in mind the underlying difficulties. It must be emphasised that good design considers all the elements, including connections, in a structure as a whole. Such an approach eliminates the problem of modifying members, already fully designed, in order to fit in connections.

To list the requirements of a good connection in steelwork is not difficult. Ideally a connection should be:

(a) rigid, to avoid fluctuating stresses which may cause fatigue failure
(b) such that there is the least possible weakening of the parts to be joined
(c) easily installed, inspected and maintained.

The two practical connection types in current use are welds and bolts. Both have aspects which are departures from the ideal characteristics listed but both also have advantages in particular circumstances.

5.2 Bolts and bolting

5.2.1 *Bolt types*

Peg fasteners of the rivet type are one of the oldest methods of joining metals; they can be traced back to man's earliest use of ductile materials. For many years the riveted joint was the preferred method of connecting steel members. However, in recent years rivets have become much less popular for a variety of reasons amongst which economics must certainly figure. Riveting of structural steelwork is now unusual and so is not considered here. But the attributes of the rivet, a connection device which at reasonable cost produced a joint which was not prone to slip when subjected to pulsating load, are now available in the high strength friction grip bolt.

Black bolts
The term 'black' is applied to unfinished common or rough bolts which have not been finished to an accurate shank dimension. Black bolts are used where slip and vibration do not matter. They are supplied in mild or higher strength steel.

Precision bolts
For connections where slip and vibration are undesirable it is necessary to use accurately machined bolts fitted into precisely drilled holes. These bolts too can be in mild steel or higher strength steels.

Both black and precision bolts are supplied to International Standards Organisation (ISO) specifications which have been adopted in the United Kingdom. The material from which the bolts are made is described by a grade classification; the two in general use are grades 4.6 and 8.8. The first figure in the grade number represents one tenth of the tensile strength of the material in kgf/mm^2, the second figure being the factor by which the first must be multiplied to give the yield stress or 0.2% proof stress in kgf/mm^2 as appropriate (see Table 5.1).

5.2.2 *High-strength friction grip (hsfg) bolts*

The replacement of site-riveting with a bolting technique which could rival the efficiency of shop-riveted joints and yet be competitive in cost is a recent

Table 5.1 Material properties of ISO bolts grades 4.6 and 8.8 (see (1) and (2))

Grade	Ultimate tensile strength (kgf/mm^2)	Ultimate tensile strength (N/mm^2)	Yield or 0.2% proof stress (kgf/mm^2)	Yield or 0.2% proof stress (N/mm^2)
4.6	40	392	24	235
8.8	80	785	64	628

innovation. In the period 1929–36 the Steel Structures Research Committee sponsored work on bolted beam to column connections. One of the products of this work was a proposal for the rivets then commonly in use to be substituted by high tensile steel bolts tightened to a controlled torque that would clamp the parts together. The idea was not pursued in Britain, however. It was left to the Americans to patent a high-strength friction grip bolt, a code for their use being issued by the American Institute of Steel Construction in 1951. The British Standard, issued in 1959, was for all practical purposes based on the American Specification.

The term 'high-strength friction grip bolts' relates to bolts of high-tensile steel, used in conjunction with high-tensile steel nuts and hardened steel washers, which are tightened to a predetermined shank tension in order that the clamping force thus provided will transfer loads in the connected members by friction between the parts and not by shear in, or bearing on, the bolts or plies of connected parts.

Because the hsfg bolt has to be installed by techniques which demand rather more expertise than merely turning a nut with a spanner some description is given here of the method of tensioning them, for unless a method is available which will guarantee a minimum shank tension the hsfg bolt will not act correctly.

Torque control method

Most national standards require that the minimum tension induced in a bolt shall be about 70% of the minimum tensile strength of the bolt. It was at first considered that the tension in the bolt shank could be accurately determined from the relationship $T = RdP_0$, where T is the applied torque, d is the nominal diameter of the bolt, P_0 is the shank tension, and R is a non-dimensional coefficient which depends on a number of factors, amongst which are the type of thread and the coefficient of friction between nut and thread.

The torque control method of tightening consists essentially of determining, by actual measurement in a suitable rig, the torque required to achieve the specific shank tension in a number of typical bolts and nuts. The bolts in the actual joint are then tightened to this torque using a hand torque wrench or a powered impact wrench calibrated to stall when the required torque has been reached. The method was incorporated in early specifications but current practice prefers the 'turn-of-nut' method described below. Although torque control gives a closer approach to the required tension in the elastic range, once the plastic range is reached thread deformations make the torque relationship erratic, and standard thread conditions are difficult to achieve in practice.

Turn-of-nut method

The bolts are first tightened sufficiently to bring the joint surfaces into close contact. This can usually be done with a hand spanner or a few blows of an impact wrench. After the preliminary tightening the nut is turned a specific

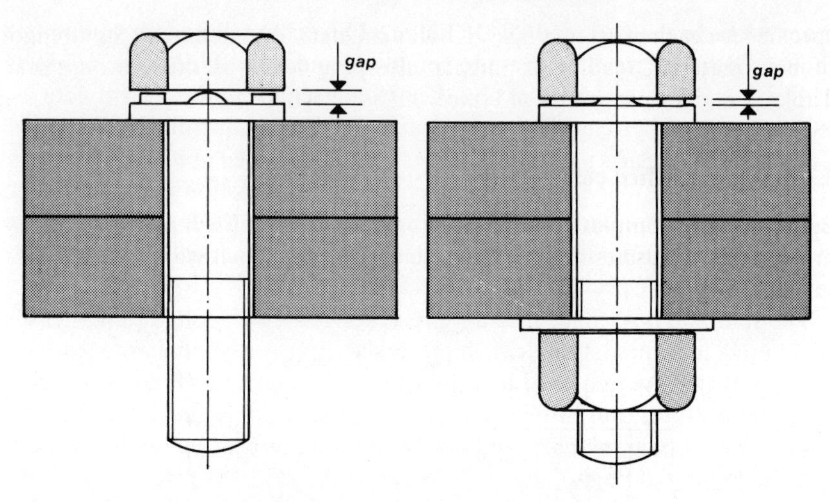

Figure 5.1 Load indicating washers
(with acknowledgement to Cooper and Turner Ltd.)

amount relative to the bolt to achieve at least the minimum shank tension; for bolts up to about 115 mm long a half-turn is sufficient. It is reassuring to know that even under the most adverse conditions at least one-and-a-half turns of the nut are required to produce failure, though a figure between two and three turns is much more usual.

There are patented methods of ensuring that the correct shank tension has been achieved by observing the deformation of the bolt head or a special washer placed under the head (see Figure 5.1).

The mechanical properties and dimensions of typical general grade bolts are given in Table 5.2*a*. Ultimate tensile and proof stresses (a lower bound to the proportional limit load and also the minimum shank tension required) are

Table 5.2a Properties of typical general grade hsfg bolts (3)

Diameter (mm)	Proof stress (N/mm^2)	Ultimate tensile stress (N/mm^2)
12–24	587	827
27–33	512	725

Table 5.2b Properties of typical higher strength hsfg bolts (3)

Diameter (mm)	Proof stress (N/mm^2)	Ultimate tensile stress (N/mm^2)
16–33	776	981

specified for each of three ranges of bolt diameters. An increase in the strength of bolt material, leading to higher clamping force, is often economical. Table 5.2*b* gives properties of typical higher strength bolts.

5.3 Design of bolted connections

Because of the complexity of connection resistance to load, design normally proceeds by the use of some realistic simplification which will produce a safe solution.

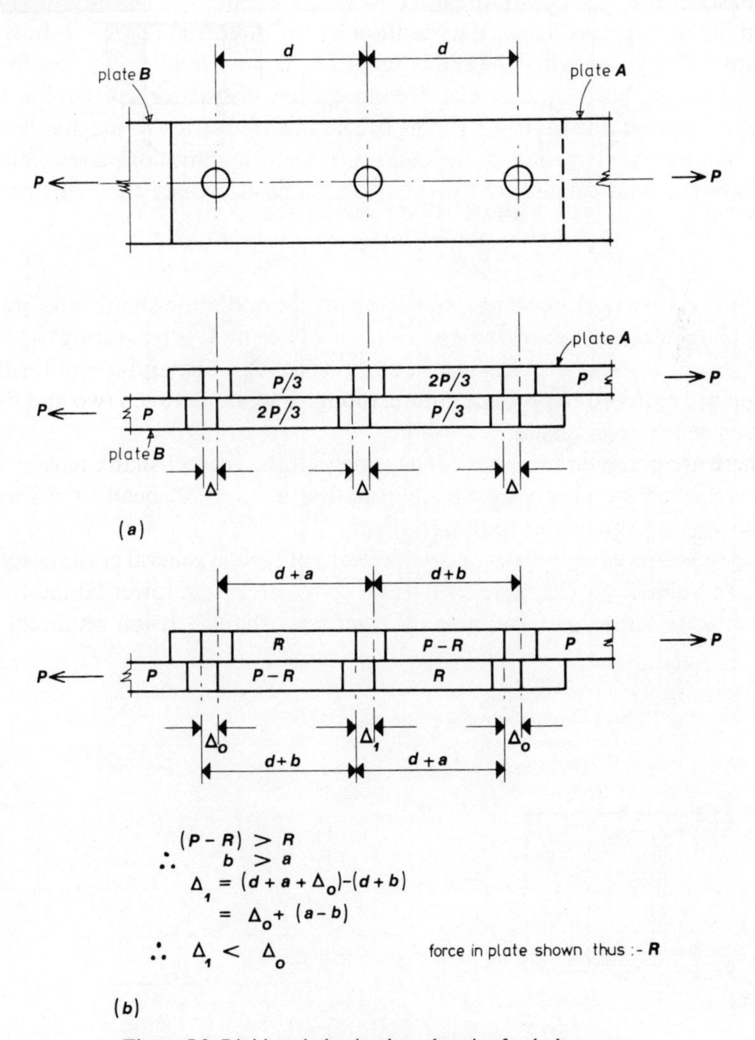

Figure 5.2 Rigid and elastic plate theories for bolt groups
(a) Rigid plate theory
(b) Elastic plate theory

5.3.1 *Load on bolts*

The distribution of the load applied to a bolted connection, between its constituent bolts, is not easily predicted. Some consideration must be given to the relative stiffnesses of the connected elements in deciding on a suitable force distribution. The lap joint shown in Figure 5.2a illustrates the problem. With the initial assumption that the plates are rigid, and the bolts elastic if plate A translates Δ with respect to plate B, all bolts are deformed equally by Δ, develop the same shearing strain and each bolt carries the same load R. This is the basis of the 'rigid plate theory' commonly adopted in design.

But if the plates are elastic the situation for the three bolted joint is shown in Figure 5.2b; by symmetry the end bolts carry the same load R. The load in the plates between bolts varies so that the elongations will also vary, that due to R being a and that due to $(P - R)$ being b. The shearing strain in the middle bolt will thus be less than that of the end bolts. Deformability of plates tends to increase the load carried by the outer bolts and decreases the loads on the

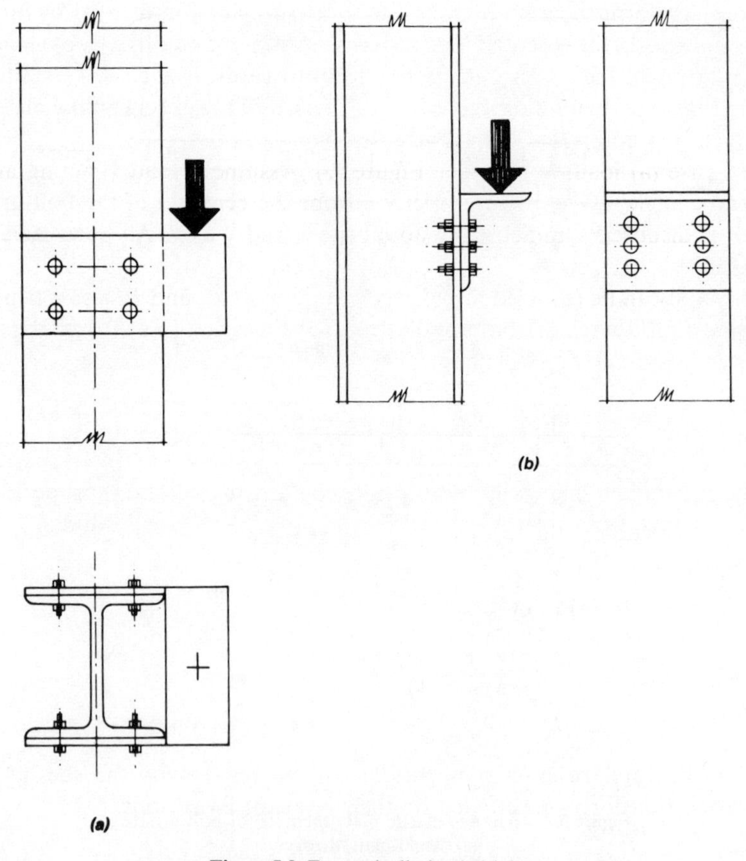

Figure 5.3 Eccentrically loaded joints

inner bolts, the effect becoming more pronounced the larger the number of rows of bolts. The elastic plate method, although seldom used in practice, should be considered as a qualitative design criterion emphasising the desirability of arranging bolts compactly in a joint so that the loads on them may be as nearly identical as possible.

In the plastic range the most heavily loaded bolts will deform without taking additional load; the loading will be redistributed to successive lines of bolts until all the bolts are stressed to the yield point. This fact, which has been experimentally verified, is the justification for using the rigid plate theory as the basis for design. The limitations of the theory as applied to very long joints must be realised; these joints may show a failure of bolts proceeding inwards, a phenomenon which has been called 'unbuttoning'.

The lap joint is a particularly simple example because the line of action of the load applied to the joint coincides with the centroid of the bolt group resisting the load. More complicated joints in which the load is eccentric to the bolt group centroid are illustrated in Figure 5.3. In case (a) the loading causes not only uniform shear on each bolt but a twisting action resisted by further shear. In case (b) the eccentric load induces tension in the bolts as well as shear. The rigid plate theory can be applied to both cases, though it is well to be aware that case (b) loading may cause distortion of the bracket if the material of which it is composed is relatively flexible.

For case (a) loading consider Figure 5.4. Assume a load P acting at an arbitrary angle α and an eccentricity e from the centroid of the bolt group which is disposed symmetrically about the x and y axes. All bolts have the same area.

The load can be resolved into direct components P_x and P_y and a twisting moment Pe. If there are n bolts in the group by the rigid plate theory the direct components on any bolt will be

$$R'_x = \frac{P_x}{n} \quad R'_y = \frac{P_y}{n}$$

The twisting moment will produce a force F on any bolt (x, y) proportional to its distance from the centroid 0. $F = kr = k(x^2 + y^2)^{\frac{1}{2}}$, or resolving

$$R''_x = F \sin \beta = F \frac{y}{(x^2 + y^2)^{\frac{1}{2}}} = ky$$

$$R''_y = F \cos \beta = F \frac{x}{(x^2 + y^2)^{\frac{1}{2}}} = kx$$

(see Figure 5.4b).

Now the total twisting moment Pe will be resisted by the sum of the individual bolt forces multiplied by their relevant lever arms

$$Pe = \Sigma(ky.y + kx.x)$$

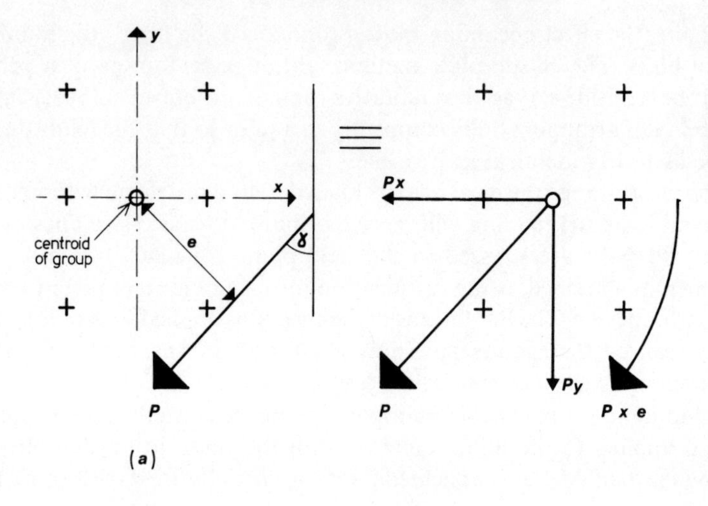

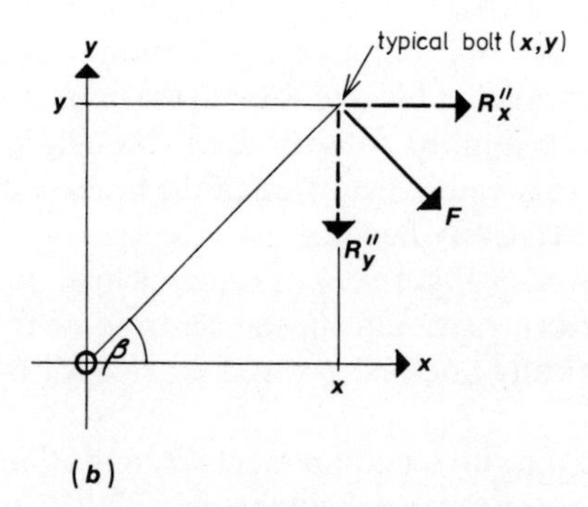

Figure 5.4 Eccentrically loaded joint analysis case (a)
(a) Equivalent load
(b) Twisting components

or

$$k = \frac{Pe}{\Sigma(x^2 + y^2)}$$

Thus

$$R''_x = Pe\frac{y}{\Sigma(x^2 + y^2)}$$

$$R''_y = Pe\frac{x}{\Sigma(x^2 + y^2)}$$

The total resultant force on any bolt can then be found from

$$\tilde{R} = [(R'_x + R''_x)^2 + (R'_y + R''_y)^2]^{\frac{1}{2}}$$

and the maximum value of $\tilde{R}$ will be the design load.

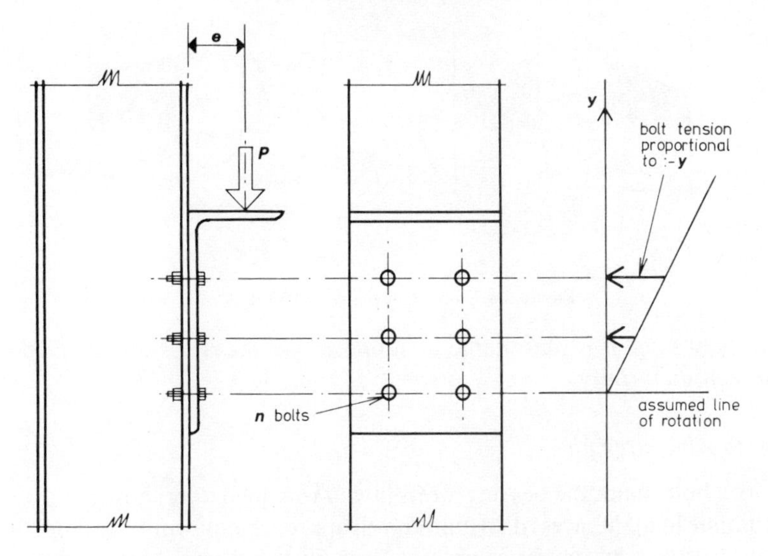

Figure 5.5 Eccentrically loaded joint analysis case (b)

Case (b) is rather more complicated because the applied load is resisted by a combination of shearing and tensile action in the bolts combined with bearing of the connected surfaces against one another. Detailed design of such connections will include considerations of local bending and buckling which are beyond the scope of this book. However, a simple approach which yields a satisfactory design basis for the bolts is to assume that the connection plate rotates rigidly about the bottom row of bolts (Figure 5.5).

In a manner similar to that for case (a) eccentric loading it can be shown for a load parallel to the y axis at an eccentricity e that vertical shear per bolt $= P/n$ and horizontal tension on bolt $(y) = Pe(y/\Sigma y^2)$. The two components of load can then be combined, using an interaction relationship, to determine the strength of bolt required (see Example 5.5).

Having considered the method of finding the load on any bolt it is necessary to investigate the local effect of this bolt load on the bolt and the adjacent plate.

5.3.2 *Stresses in plates*

The presence of holes in a plate under tension leads to stress concentrations which may be severe enough to produce local yielding at working loads. If the load is increased, however, the distribution becomes more nearly uniform, plastic behaviour leading to a smoothing-out of stress distribution (Figure 5.6).

5.3.3 *Shearing strength*

It is not easy to determine the actual distribution of shearing stress in a bolt in the elastic range; it is certainly not uniform. But at ultimate load, because of the

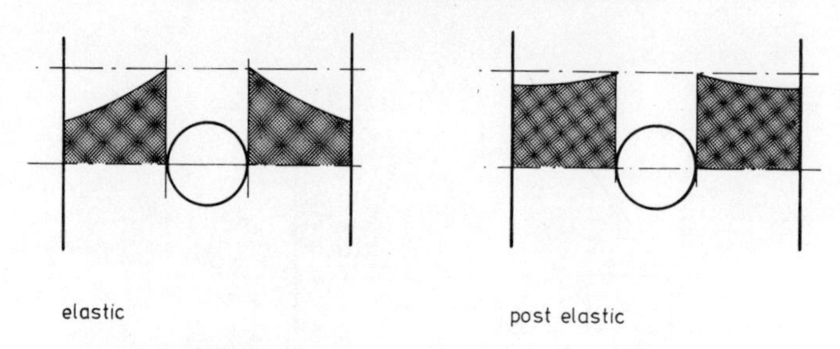

Figure 5.6 Stresses in plates across holes

ductility of steel, it is reasonable to suppose that the stress distribution will approach uniformity.

5.3.4 *Bearing strength*

Across a bolt shank the bearing stress at working load may show high peaks. At ultimate load the stress distribution will approach uniformity, though there are no satisfactory measurements of these distributions. Along a bolt the bearing stress distribution is dependent to some extent on whether the bolt is being bent. For a portion of a bolt between two plates the distribution is nearly uniform but there may be some variation in the side plates. A single lap joint shows considerable variation (Figure 5.7). For this reason, where bolts are in single shear, permissible bearing stresses may be reduced under some code of practice rules.

5.3.5 *Tensile strength*

Tensile failure normally occurs in the threaded portion of a bolt at the root of the thread over the tensile stress area.

5.3.6 *Bolt capacities*

Shear

Effective area: If threads do not occur in the shear plane the shear area A_s may be taken as the shank area A, otherwise as the tensile stress area (the area at the bottom of the threads) A_t.

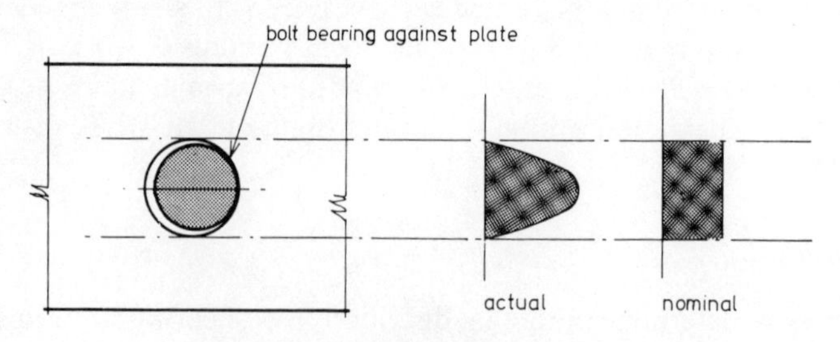

Figure 5.7 Bearing stresses

In the absence of reduction for long joints or large grips the shear capacity $P_s = p_s A_s$, where p_s is the bolt shear strength. A *long joint* is one, in a splice or end connection in a tension or compression element containing more than two bolts, in which the distance L_j between the first and last rows of bolts, measured in the direction in which the load is transferred, exceeds 500 mm. In such a joint P_s is reduced by a factor $(5500 - L_j)/5000$ to $P_s = p_s A_s (5500 - L_j)/5000$.

A *large grip length* is one in which the total thickness of the connected plies T_g exceeds five times the nominal diameter d of the bolts. In such a joint P_s is reduced by a factor $8d/(3d + T_g)$ to $P_s = p_s A_s \times 8d/(3d + T_g)$.

Where a joint is both *long* and has a *large grip length* the lower value of P_s is to be used.

Bearing capacity
The effective capacity is the lesser of the bearing capacity of bolt, P_{bb}, and connected ply P_{bs}. For the bolt $P_{bb} = dtp_{bb}$, where d is the nominal diameter of the bolt, t is the thickness of the connected ply and p_{bb} is the bolt bearing strength. For the connected ply $P_{bs} = dtp_{bs} \leqslant 0.5\ etp_{bs}$. The latter expression takes account of the increased risk of failure where the end distance e falls below approximately two bolt diameters.

Tension capacity
The tension capacity, $P_t = p_t A_t$ where p_t is the tension strength.

The various strength values are summarised in Table 5.3. The calculation of bolt capacities is shown in Example 5.1.

5.3.7 *Interaction between tension and shear*

In one type of eccentric connection bolts are loaded in combined shear F_s and tension F_t. For this combination the interaction formula $(F_s/P_s) + (F_t/P_t) \leqslant 1.4$ has been adopted.

It will be apparent that, for instance, although F_s must never exceed its maximum allowable value P_s, there is, even at this stage, some tension capacity still left in the bolt: $(F_t/P_t) = 1.4 - 1.0 = 0.4$ (when $F_s = P_s$) and similarly $(F_s/P_s) = 1.4 - 1.0 = 0.4$ (when $F_t = P_t$).

This interaction formula is shown graphically in Figure 5.8.

Table 5.3 Summary of ordinary bolt strengths

Shear capacity	$P_s = p_s A_s d$
Bearing capacity	
bolt	$P_{bb} = dtp_{bb}$
ply	$P_{bs} = dtp_{bs} \leqslant 0.5\ etp_{bs}$
Long joints	$P_s = p_s A_s[(5500 - L_j)/5000]$
Large grips	$P_s = p_s A_s[8d(3d + T_g)]$
Tension capacity	$P_t = p_t A_t$
Shear and tension	$(F_s/P_s) + (F_t/P_t) \leqslant 1.4$

BS5950 clause	Example 5.1 Bolt capacities	5.1.1
6.3	<u>M20 black bolt, grade 4.6, to BS4190</u>	
	nominal diameter d	20 mm
	shank area	314 mm^2
	tensile stress area A_t	245 mm^2
Table 32	tension strength p_t	195 N/mm^2
Table 32	shear strength p_s	160 N/mm^2
Table 32	bearing strength p_{bb}	435 N/mm^2
6.3.6.1	<u>Tension capacity</u>	
	$P_t = 245 \times 195 \times 10^{-3}$	= **47.8 kN**
6.3.2	<u>Shear capacity</u>	
	threads in shear plane	
	$P_s = 245 \times 160 \times 10^{-3}$	= **39.2 kN**
6.3.1	no threads in shear plane	
	$P_s = 314 \times 160 \times 10^{-3}$	= **50.2 kN**
6.3.3	<u>Bearing capacity in 6mm grade 43 plate</u>	
6.3.3.2	$P_{bb} = 435 \times 20 \times 6 \times 10^{-3}$	= **52.2 kN**
Table 33	$p_{bs} = 460$ N/mm^2	
6.3.3.3	bolt governs bearing capacity unless	
	end distance $e < 2 \times (p_{bb}/p_{bs})\, d$	
	$< 1.89\, d$	

BS5950 clause	Example 5.1 Bolt capacities 5.1.2
6.4	M20 high strength friction grip bolt.
	nominal diameter d 20 mm
	minimum shank tension P_0 144 kN
6.4.4.2	Tension capacity
	$P_t = 0{\cdot}9 \times 144 = \boxed{130\,kN}$
6.4.2.1	Slip resistance
	$P_{SL} = 1{\cdot}1 \times 0{\cdot}45 \times 144 = \boxed{71{\cdot}3\,kN}$ $[K_s = 1{\cdot}0]$
6.4.2.2	Bearing resistance in 6mm plate
Table 34	$p_{bg} = 825\ N/mm^2$ (grade 43)
	$P_{bg} = 20 \times 6 \times 825 \times 10^{-3} = \boxed{99\,kN}$
	(end distance $\geq 3 \times 20 = 60mm$)

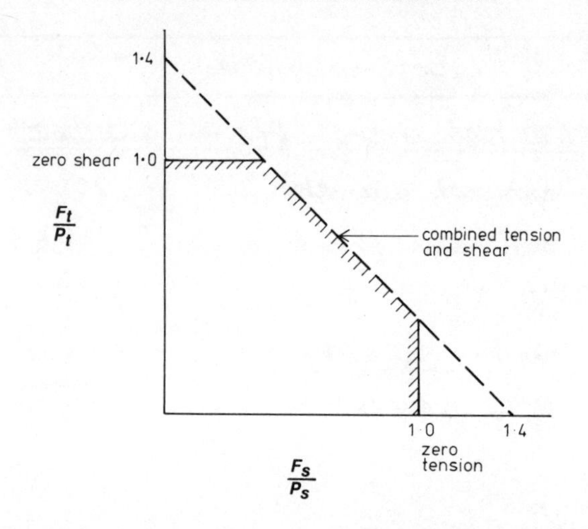

Figure 5.8 Interaction diagram for tension and shear in black bolts

5.3.8 *Spacing*

As well as avoiding bolt failure it is necessary to formulate rules to prevent

(a) tear out failure of a bolt through the edge of the plate (Figure 5.9a)
(b) splitting of the plate between bolts (Figure 5.9b)
(c) buckling of the plate between bolts or at the edge (Figure 5.10).

Minimum and maximum end and edge distances are specified to avoid tear out and distortion respectively. Minimum and maximum spacing restrictions are needed to cope with splitting and buckling. Where members are exposed to corrosive conditions further restrictions apply so that joints are tightly sealed. Table 5.4 summarises these rules.

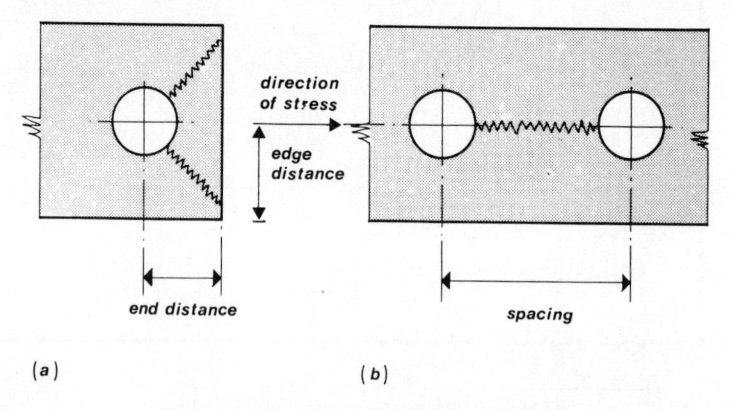

Figure 5.9 Edge and end distances and spacing

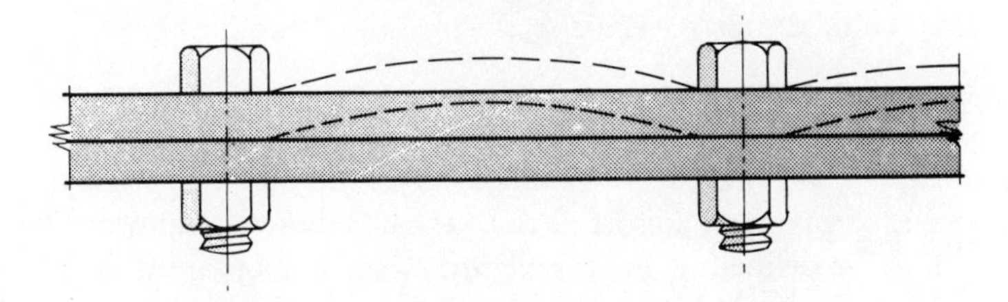

Figure 5.10 Distortion of plate with widely spaced bolts

Table 5.4 Fastener spacing and edge distances

	Required distance*
Minimum spacing	$2.5d$
Maximum spacing	
normal (in direction of stress)	$14t$
exposed (in any direction)	$16t \ngtr 200$ mm
Minimum edge and end distance	
rolled, machine flame cut	
or planed edge	$1.25D$
sheared or hand flame cut edge	$1.4D$
any end	$1.4D$
Maximum edge distance	
normal	$11t\varepsilon$
exposed	$40\text{ mm} + 4t$

* where t is the thickness of the thinner part, d is the nominal bolt diameter, D is the hole diameter and $\varepsilon = (275/p_y)^{1/2}$.

Bolts are often staggered in order to produce as compact a joint as possible and in this case two terms need to be defined:

pitch which is the distance between adjacent bolts in a line parallel to the direction of stress in the member,
gauge which is the perpendicular distance between two consecutive lines of bolts parallel to the direction of stress (Figure 5.11).

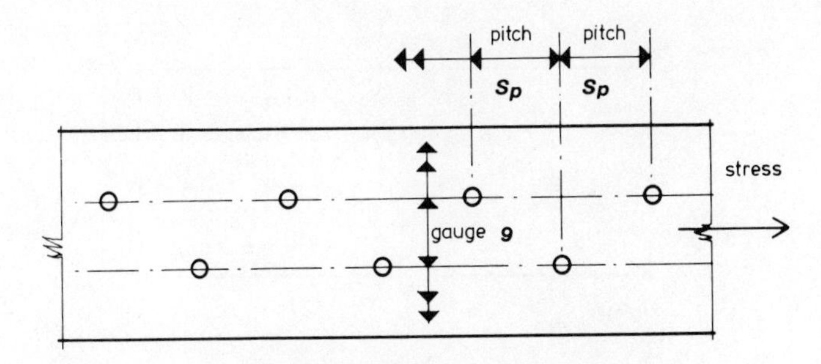

Figure 5.11 Pitch and gauge

5.3.9 *Net area of tension members*

A point which requires attention is the mode of failure of joints in tension members in which the bolts are staggered. The problem of net section has generated a very large amount of study but as yet there does not seem to be any conclusive experimental support of the various theoretical approaches. The problem may be defined as the prediction of the line of failure in a tension member containing holes. In some cases the answer is obvious (Figure 5.12a) but for the arrangement in Figure 5.12b it is certainly not clear which of the possible failure modes shown will actually occur although there must be a location of hole E which will make the breaking of the plate equally probable along the transverse and zig-zag lines. Attempts to show this mathematically have not proved successful.

A commonly adopted approach is based on the work of Cochrane (4). The basic net section is taken as that right section giving the smallest net area.

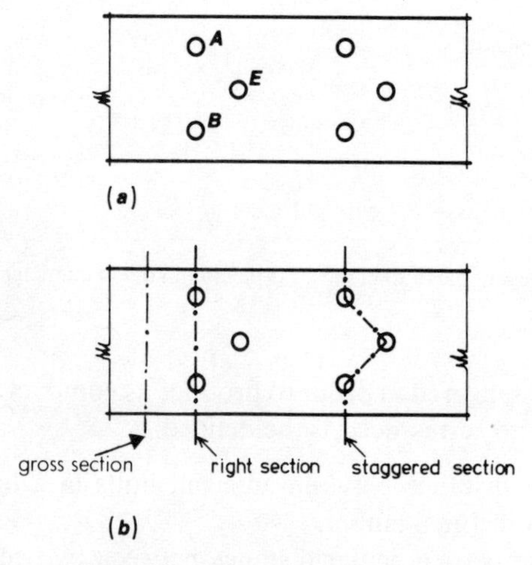

Figure 5.12 Failure of plate with holes

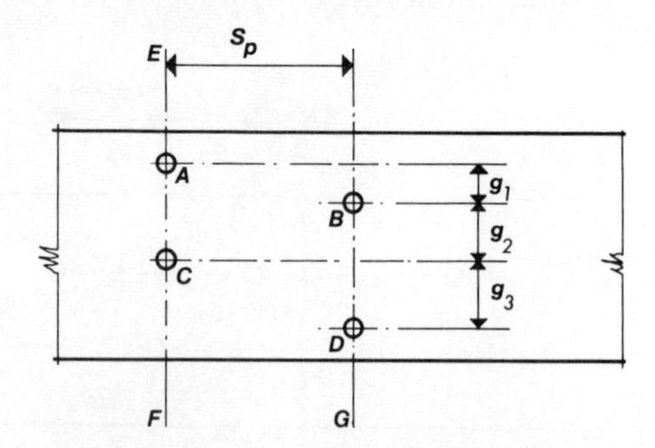

Figure 5.13 Staggered section

Staggered holes are then given consideration by deducting a certain portion of each staggered hole from the gross right section, according to the formula: $x = 1 - (S_p^2/4gd)$, where x is the portion of the staggered hole to be deducted from the gross section, S_p is the pitch measured parallel to the stress, g is the gauge and d is the diameter of hole. Note that S_p and g are measured for each hole from the one preceding it. For example, consider the arrangement in Figure 5.13. The net right section *EF* contains two holes. The number of holes to be deducted on the section *EABCDG* is $n = 1 + x_b + x_c + x_d$ and this will rule if n is greater than 2. For given values of S_p, d and g it is possible to plot curves as in Figure 5.14 from which values of x may be read. The labour of finding net sections along a number of possible lines is then substantially reduced. Another possibility is deliberately to arrange the value of pitch such that $S_p^2/4gd \geqslant 1$. Then x will be negative or zero and the staggered section will automatically be larger than the net right section (see Example 5.2).

5.3.10 *Friction grip fasteners*

Restricting consideration to the ordinary parallel shank bolt in a clearance hole the transverse capacity is the smaller of the slip resistance (modified if necessary for long joints) and the bearing resistance.

Slip resistance

For a slip factor (coefficient of friction) μ and a bolt shank tension P_0 the minimum shearing force to cause a two plate connection to slip is $P_{sL} = \mu P_0$. Slip is a serviceability condition but it is possible to alter the expression so that a check on the ultimate shear capacity of the bolt shank after slip has occurred and the bolt is bearing against the plates is not necessary.

Assuming a practical upper limit of 0.55 for the slip factor, ultimate shear

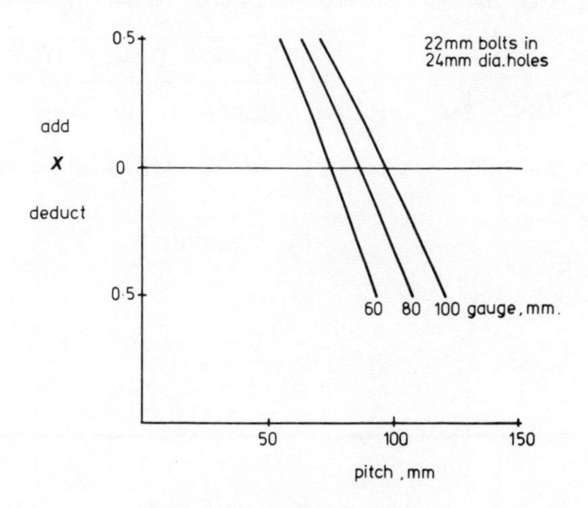

Figure 5.14 Net section graph

BS5950 clause	Example 5.2 Net area of a tension member 5.2.1
3.4	
	M20 bolts in 22 mm dia holes, 20 mm thick plate.
	Gross plate area $300 \times 20 = 6000\ mm^2$
3.4.2	Minimum sectional area at right angles is on line 1-1 = gross area − 3 holes
3.4.3.	Staggered section ABCDEFG
	$= \text{gross area} - 1 - x_C - x_D - x_E - x_F$
	where $x_C = x_D = x_E = x_F = 1 - (S_p^2/4gD)$
	$= 1 - (60^2/4 \times 50 \times 22) = 0{\cdot}18$
	staggered area = gross area − $(1 + 4 \times 0{\cdot}18)$ holes
	= gross area − 1·72 holes
	Thus the ruling section is 1-1 having a net area of $6000 - 3(22 \times 20) = 4680\ mm^2$

resistance $= 0.6P_0$; $0.6P_0/0.55P_0 = 1.1$ therefore if the slip resistance is calculated from $P_{sL} = 1.1\mu P_0$ and checked against factored ultimate load a further check on the shank shear resistance is not needed.

Bearing resistance
The bearing capacity is obtained from $P_{bg} = dtp_{bg} \leqslant 1/3etp_{bg}$. Note the increased importance of the end distance e compared with that for ordinary bolts.

Long joints
Long joints are defined in the same way as for ordinary bolted connections and the slip resistance is given by $P_{sL} = 0.6P_0[(5500 - L_j)/5000]$.

Tension capacity
The effect of applied tension is in fact very small provided the plates do not separate. If a bolt grips two plates (Figure 5.15) and a load is applied to the bolt it will stretch, but provided the load is not greater than the initial bolt tension the two plates will remain in contact. The bolt elongation and the plate expansion must be equal. For bolts and plates of the same material, under elastic conditions, the increase in bolt tension will depend on the relative stiffness of bolt and plate.

The increase in bolt tension ΔP_0 may be written

$$\Delta P_0 = \frac{F}{1 + A_p/A_b}$$

where A_p is the area of plate compressed and A_b is the area of bolt and F is the applied tensile load. It is not possible to give an exact value to A_p but a lower limit might be $10A_b$. Assuming that the applied load $F = 0.5P_0$, where P_0

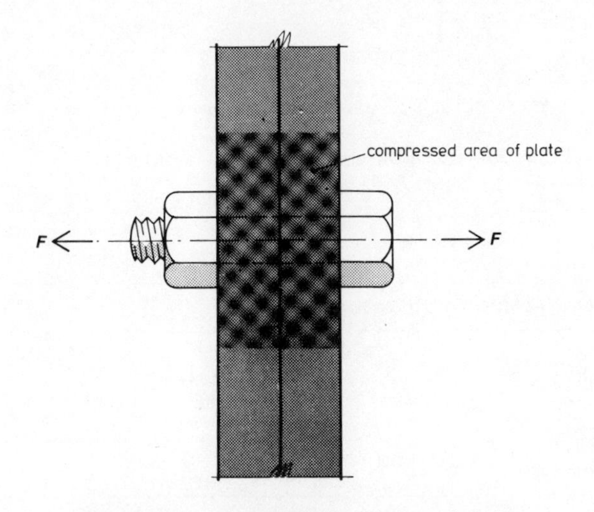

Figure 5.15 Tension applied to friction grip bolt

is the initial tension in the bolt, and that $A_p = 10A_b$ the final tension in the bolt will be $P_0 + 0.045P_0 = 1.045P_0$. It is not until F becomes so great as to cause plate separation that the bolt takes all the load.

The tension capacity P_t is calculated from $P_t = 0.9P_0$.

Combined shear and tension

Subject to the fastener having adequate capacity in shear and tension considered separately it should also satisfy $(F_s/P_{sL}) + 0.8(F_t/P_t) \leqslant 1.0$, where F_s is the applied shear, and F_t is the external applied tension. This interaction relationship is illustrated in Figure 5.16. The various strength values are summarised in Table 5.5.

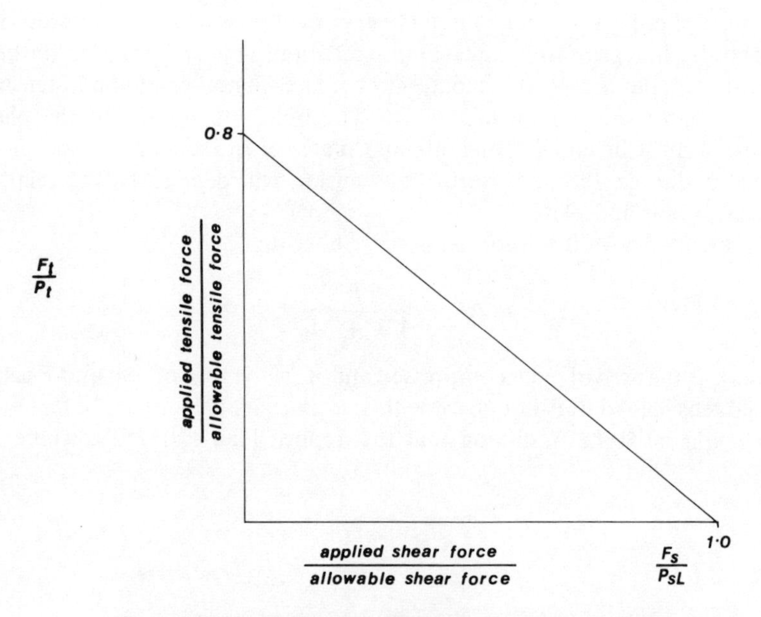

Figure 5.16 Combined shear and tension in friction grip bolts

Table 5.5 Summary of friction grip bolt strengths

Slip resistance	$P_{sl} = 1.1\ K_s \mu P_0^*$
Bearing capacity	$P_{bg} = dtp_{bg} \leqslant 0.33\ etp_{bg}$
Long joints	$P_{sl} = 0.6P_0[(5500 - L_j)/5000]$
Tension capacity	$P_t = 0.9P_0$
Shear and tension	$(F_s/P_{sl}) + 0.8(F_t/P_t) \leqslant 1.0$

$^*K_s = 1.0$ for fasteners in clearance holes

5.4 Design examples

The process of designing a bolted joint requires the following sequence:

(a) make an initial estimate of the number and layout of bolts required
(b) calculate the maximum bolt load
(c) select a bolt suitable to carry the load
(d) check the strength of critical plate sections
(e) check bolt spacing and edge distances for compliance with code rules.

Design is very much facilitated by the use of tables of bolt capacity. Full tables are given in handbooks (5); their manner of construction is illustrated in Example 5.1.

Example 5.3. Bolt group loaded through its centroidal axis in shear
The joint is a butt splice in a tension member. Because the bolts are symmetrically placed about the longitudinal axis of the member they are loaded only in shear. The number of bolts required is rapidly estimated by consideration of the smaller of the shearing or bearing strength of a given bolt size. The net section of both main and cover plates must be checked to ensure that the allowable tensile stress in the plate material is not exceeded.

Example 5.4. Bolt groups eccentrically loaded in shear
The load, being eccentric to the centroid of the bolt group, produces two components of shear force; a direct component shared equally by all bolts, and a twisting component whose action on any bolt is proportional to that bolt's distance from the centroid of the group. An arrangement of bolts having been made, the resultant force on the most heavily loaded bolt can be calculated and a suitable bolt size selected.

Example 5.5. Bolt group eccentrically loaded, producing shear and tension
The load on this group is equivalent to a direct shear component shared equally by all bolts, and a twisting component. The latter may be considered either resisted wholly by the tensions in the bolts, being proportional to the distance of the bolt from some assumed line of rotation, or by a combination of bolt tensions and compression in the connection plate. Once an arrangement of bolts has been made the tensile force on the most heavily loaded bolt can be calculated and a suitable bolt size selected. Tables of coefficients can be used to simplify the work.

The compressed depth of the connection plate is difficult to quantify but the capacity of the joint is not greatly affected by quite large changes in the assumed depth of compression.

The capacity of the worst loaded bolt is restricted by the interaction formula

BS5950 clause	Example 5.3 Butt splice in a tension member 5.3.1
	Butt splice 20 mm thick, 150 mm wide plates carrying a factored axial tensile load of 500 kN.
Table 6	Grade 43 steel, p_y = 265 N/mm²
	Trial layout:
	A B C D 150 12 20 12
	M20 grade 4.6 bolts in 22 mm holes
	double shear capacity 2× 39.2 = 78.4 kN
	bearing in 20 mm plate > 78.4 kN
	bolt capacity = 78.4 kN
	number of bolts 500/78.4 say 7
	Check capacity of net section of main and cover plates at rows A B C D.
	Gross areas
	main plate 150× 20 = 3000 mm²
	cover plate 150 × 12 = 1800 mm²

BS5950 clause	Example 5.3 Butt splice in a tension member 5.3.2

Main plate

		row A	B	C	D
3.3.3	Effective area A_e mm²	3000	2544	2544	2544
	Capacity kN	795	674	674	674
	Load kN	500	429	286	143

Cover plate

		A	B	C	D
	Effective area A_e mm²	1800	1526	1526	1526
	Capacity kN	477	404	404	404
	Load kN	36	107	179	250

3.3.3 [note - K_e for grade 43 steel is 1·2 - effective area = 1·2x net area ≯ gross area]

Bolt layout

d = 20 mm D = 22 mm

Clause			
6.2.1	minimum spacing	2·5×20	= 50 mm
6.2.3	minimum edge distance, rolled edge:	1·25×22	= 28 mm
6.2.3	end distance say	2×20	= 40 mm
6.2.4	maximum edge distance	11×12ε	≈ 130 mm
6.2.2	maximum spacing	14×12	= 168 mm

BS5950 clause	Example 5.3 Butt splice in a tension member 5.3.3
	Redesign using 20mm hsfg bolts
	Minimum shank tension P_0 = 144 kN (BS4604)
	μ = 0·45
6.4.2.1	P_{SL} = 1·1 × 0·45 × 144 = 71·3 kN
	For end distance of 40mm
6.4.2.2	$P_{bg} = \frac{1}{3} 40 \times 20 \times 825 \times 10^{-3}$ = 220 kN
	Ruling capacity is double shear:
	2 × 71·3 = 143 kN
	number of bolts 500/143 say 4
	[from previous calculation main plate and cover plates with 2 holes have adequate capacity]

BS5950 clause	Example 5.4 Bolted column bracket 5.4.1

Grade 43 steel

16mm plates

305x305x97 UC

T = 15.4mm

300

2P kN

A

75 75 75 75

70 70

x y

$\Sigma x^2 = 10 \times 70^2 = 49 \times 10^3$

$\Sigma y^2 = 4 \times 75^2 = 22.5 \times 10^3$

$4 \times 150^2 = 90 \times 10^3$

$\Sigma(x^2 + y^2) = 162 \times 10^3\ mm^2$

$= \boxed{0.162\ m^2}$

Forces on bolt A:

$R'_x = 0, \quad R'_y = P/10 = 0.100P$

$R''_x = 0.3P(0.15/0.162) = 0.278P$

$R''_y = 0.3P(0.07/0.162 = 0.130P$

Total x component $0.278P$

Total y component $(0.10 + 0.13)P = 0.23P$

resultant $(0.278^2 + 0.23^2)^{1/2}P = \boxed{0.361P\ kN}$

M20 black bolts, grade 4.6:

single shear 39.2 kN, bearing in 15.4mm > 39.2 kN

$P_{max} = 39.2/0.361 = \boxed{109\ kN}$

BS5950 clause	Example 5.5 *Bolted end plate*	5.5.1

190kN

30kNm

10mm end plate

6.8mm flange

60

220

A- -A

46 60 46

M20 grade 4.6 bolts:

P_s shear capacity 39.2 kN

P_t tension capacity 47.8 kN [example 5.1.1]

Assume rigid plate rotation about A-A:

$\Sigma y^2 = 2(220^2 + 280^2) = 254 \times 10^3 \text{mm}^2$

$T_2 = 30 \times 10^3 (280 / 254 \times 10^3) = 33\text{kN}$

$= 33/2 = 16.5 \text{kN/bolt}$

T_2 T_1 280 220 A-A

$F_s = 190/6 = 31.7 \text{kN}$

$F_t = 33 \text{kN}$

4.3.6.3 $F_s/P_s = 31.7/39.2 = 0.808$

$F_t/P_t = 16.5/47.8 = 0.345$

$1.153 < 1.4$

$(F_s/P_s) + (F_t/P_t) \not> 1.4$. It should be kept in mind that if, as is normally the case, the threaded portion of the bolt shank is not in the shear plane then at this plane F_s is calculated on the full shank area and not on the net area at the thread root.

5.5 Welded connections

Welding is a process of connecting pieces of metal together by application of heat with or without pressure; the process is of great age and in primitive form must have been known to Iron Age men. The term 'welding' covers a large number of different processes but for welding structural steelwork the most commonly used technique is the electric arc process in which the pieces of metal to be joined are fused together by an electric arc, additional metal being added at the same time from a rod or wire. The heat generated by the arc is sufficient to join the metals without pressure. Early arc welds (the process dates from 1881) were made with carbon electrodes, but metal electrodes were soon substituted. The new art made slow progress until after the First World War. Between the wars and during the Second World War arc welding became much more widely used; in fact the use of welding was often ahead of the development of safe welding procedures and this led, in particular, to the problem of brittle fracture (see Section 1.7).

5.5.1 *The welding process*

In the electric arc welding process heat is generated by an electric arc formed between the electrode and the parts to be joined. The arc temperature, which has been estimated to be as high as 6000 °C, is sufficient not only to melt the electrode but also the metal in the joint face and this leads to a desirable penetration of the electrode metal into the joint. A problem in early welding practice using bare metal electrodes was embrittlement of the completed joint by absorption of atmospheric nitrogen. Modern electrodes are coated with a flux which melts and vaporises under heat, producing a gaseous shield which stabilizes the arc and prevents the molten metal from absorbing atmospheric gases. At the same time the molten coating makes a slag which rises to the top of the molten metal and there forms a further shield from the atmosphere.

Manual welding is slow and requires considerable skill. Whenever possible, therefore, automatic welding using long lengths of bare electrode wire and an independent source of protection to the molten metal is employed. In the submerged arc process, powdered flux is added from a separate source. Increasing use is also being made of inert gases (carbon dioxide or argon) as shields to the molten metal in both manual and automatic welding equipments.

A proper appreciation of the effect of welding on material properties is important; ignorance of this has been the cause of a number of failures in welded structures. A satisfactory weld requires a judicious choice of material and electrode; the choice of the latter may not be the responsibility of the designer but he must have some knowledge of welding metallurgy. When steel is heated above a critical temperature of about 815 °C, its structure is almost uniform, crystalline austenite. Fast cooling from this condition produces a brittle structure containing much martensite, whereas slow cooling produces a ductile, pearlite structure. The actual cooling rate and critical temperature vary with the composition of steel and in particular with its carbon content. Thick sections, which provide a large mass of material and thus absorb heat rapidly, will cool quickly, tending to become brittle. To avoid this problem such sections will require preheating by raising the temperature of the member before welding so that heat flow from the weld area is reduced. Particular care is always necessary with high-carbon steels. The heat input from a weld may be increased by increasing the size of the weld.

The position of the electrode in relation to the work is important since it affects the quality and cost of the weld. The best position is the downhand or flat, the worst is overhead; overhead welds are possible because the magnetic field in the arc carries the molten metal on to the joint. Considerable trouble is taken in fabrication shops, by the use of positioning devices, to make welds in the downhand position and designers should always be aware of the penalties in cost which may be incurred through inconsiderate weld detailing (6).

5.5.2 *Types of welded connection*

The two basic types of weld are the fillet and the butt weld. Some variations of these are shown in Figure 5.17.

Butt welds

In order that the weld may develop the full strength of the members joined, it is necessary to ensure full penetration of weld metal into the joint. Unless such penetration can be guaranteed the joint is known as an incomplete penetration butt weld. The amount of preparation required to attain full penetration depends on the thickness of plate to be joined, ranging from no preparation on thin plate up to 3 mm thick to double V or double J preparation for plate over 37 mm thick (Figure 5.18). Shaping may be done by flame cutting or machining. The finished weld is normally convex, a fact which leads to undesirable stress concentrations. For the best class of work (having the highest fatigue resistance) the weld is ground off flush. Similar considerations apply to butt welds joining plates of unequal thickness.

Fillet welds

The standard weld has equal leg lengths and a flat or convex face. It is possible to have a concave face; the effect of this on the nominal weld size is shown in Figure 5.19. As the beginning (start) and end (stop) of a fillet weld are not effective in resisting stress it is normal practice to specify that fillets should be

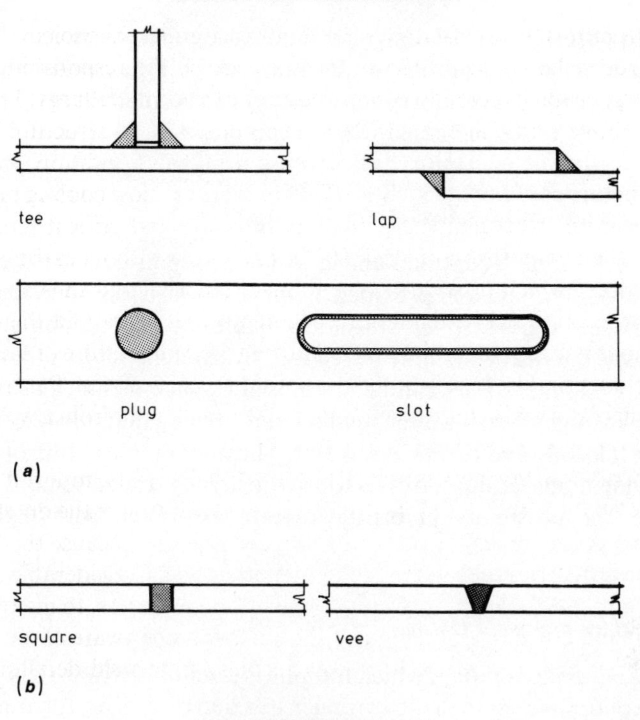

Figure 5.17 Types of weld
(a) Fillet welds
(b) Butt welds

Figure 5.18 Edge preparation for butt welds

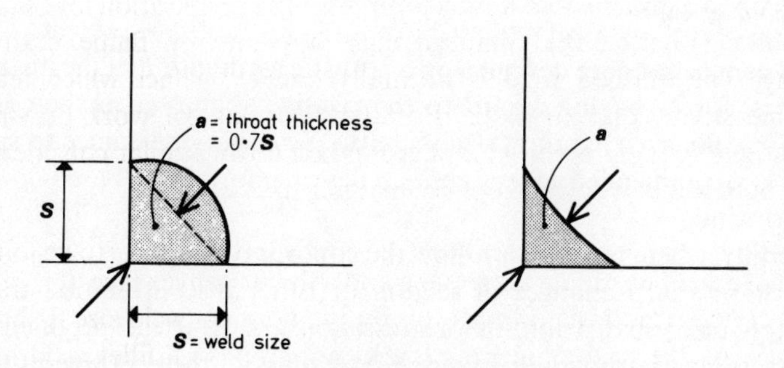

Figure 5.19 Fillet weld sizes

continued past the point of design for a distance equal to twice the size of the weld; these extensions are known as *end returns*. Fillet welds need not be continuous; subject to certain conditions they can be intermittent. The amount of weld metal can be increased where necessary with plug or slot welds.

5.5.3 *Distortion and residual stress*

As weld metal cools it contracts and in so doing may distort the plates to which it is attached. The restraint provided by the plates leads to the setting up of internal residual stress systems. It is not possible entirely to eliminate either of these effects but distortion may be reduced by careful attention to sequence of welding. Residual stresses can be minimised by such methods as preheating, stress relieving by heating after welding and peening (hammering the weld to elongate it locally and relieve shrinkage). The undesirable nature of distortion is self-evident but residual stresses too are generally troublesome. There is, for example, the possibility of brittle fracture associated with high residual stresses.

5.5.4 *Welder and weld testing*

Welding is a skilled craft for which training is essential. Because it is possible to disguise poor welding so that by visual inspection there is no apparent defect, it is common practice to require a welder to pass tests before he is permitted to work on a welded structure. In this way there will be the best possible assurance that welds will be well made.

Nevertheless, some unavoidable defects may still be present which will require detection. A variety of methods is in use for either a random or complete investigation of welds. Where weld failure would be disastrous (as, for example, in a main butt weld in the tension flange of a bridge girder) then the whole weld will be tested by radiography. For less important welds simpler methods using ultrasonics or dyes are available.

5.5.5 *Weld strength*

Welded connections are designed on a realistic assumption of the distribution of internal forces having regard to the relative stiffnesses of the connected parts. To make such assumptions ab initio is not easy; reference to specialist texts (7) is recommended as they contain design information based on analysis of experiments.

Generally, where butt welds follow the contour of the parts to be joined and do not have sharp changes in section or other discontinuities, there is a reasonably uniform distribution of stress. Some of the effects producing stress concentrations are shown in Figure 5.20. To avoid these problems, welding specifications contain requirements for the grinding of weld faces, tapering of

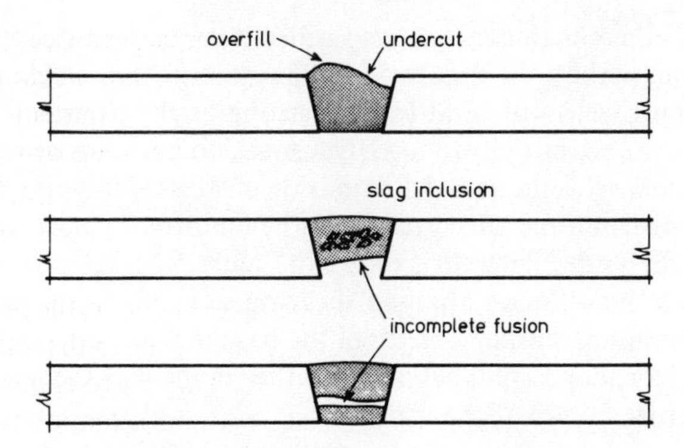

Figure 5.20 Some defects in butt welds

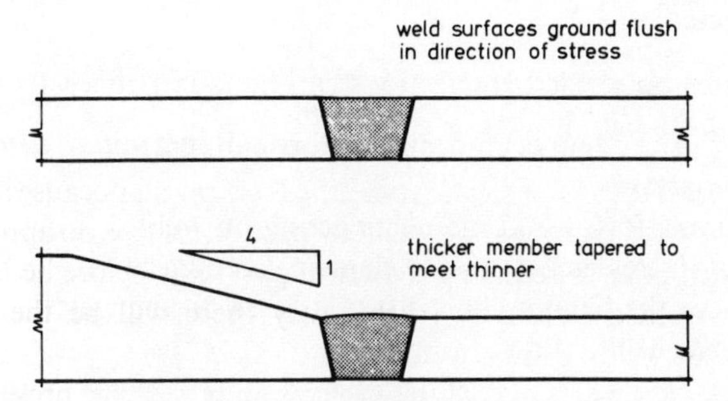

Figure 5.21 Good practice in butt welds

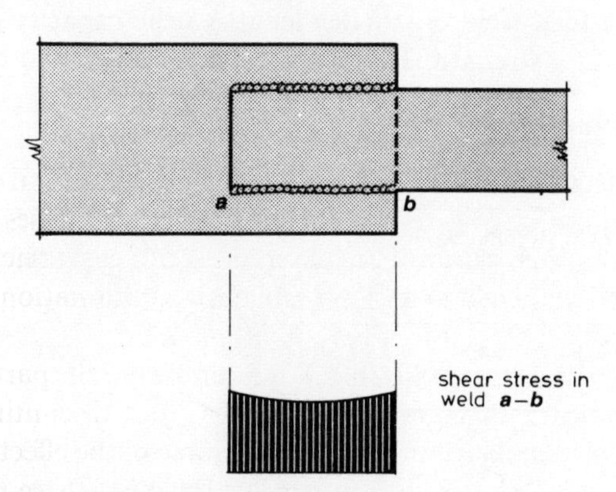

Figure 5.22 Stress in fillet weld

members of different thicknesses and other items necessary to avoid stress concentrations in the weld (Figure 5.21.) If these requirements are met it can be said that a butt weld will be at least as strong as the parts it joins.

Because of the eccentricity of force transmission and discontinuity of shape inherent in fillet welds the actual distribution of stress is complex. In the elastic range stress distribution will certainly not be uniform. Typical working stress distribution for a side fillet weld is shown in Figure 5.22. Tests have shown that for side fillets the ultimate nominal shearing strength is not less than two-thirds the minimum tensile strength of the base metal. Long, large-size fillets are weaker than short, small-size fillets; rather in the way that long rivet lines show premature failure at the ends.

The assumptions made in the design of welded connection are:

(a) the welds are homogeneous, isotropic elastic elements
(b) the parts connected by welds are rigid
(c) effects of residual stresses, stress concentrations and shape of weld are neglected

The stresses on welded connections can be found generally as follows:

(a) establish a system of co-ordinates through centroid of weld; the axes are principal axes
(b) determine forces and moments acting on weld
(c) the total stress acting on any element can then be determined by adding together the bending and direct stresses on the weld.

5.5.6 *Capacity of fillet welds*

The design strength of a fillet weld p_w is related to the strength of the lowest grade of material joined in accordance with Table 5.5.

It is convenient when making calculations to use the capacity per unit length of weld. For a fillet weld of size (leg length) S the capacity per unit length is given by $(S/\sqrt{2}) \times$ strength. Table 5.6 has been calculated on this basis.

Table 5.5 Fillet weld strength p_w

Grade to BS4360	Electrode to BS639	
	E43	E51
	p_w(N/mm^2)	
43	215	215
50	215	255

Table 5.6 Capacity of fillet welds

Weld size (mm)	Throat (mm)	Capacity (kN/mm) Grade 43 or 50
3	2.1	0.452
4	2.8	0.602
5	3.5	0.753
6	4.2	0.903
8	5.6	1.20
10	7.0	1.51
12	8.4	1.81
15	10.5	2.26
18	12.6	2.71
20	14.0	3.01
22	15.4	3.31
25	17.5	3.76

5.6 Design examples

Example 5.6 Weld eccentrically loaded in its own plane

The bracket of Example 5.4 is redesigned for fillet welding. Eccentric load produces two components of shear force at any point in the weld. These components are added vectorially and the maximum resultant used as the weld design value.

Example 5.7 Weld eccentrically loaded out of its own plane

The connection of Example 5.5 is redesigned for fillet welding. The beam is welded directly to the column by fillet welds. Eccentric load produces a shear force in the plane of the weld and a bending force at right angles to the weld plane. These components are added vectorially and the resultant, considered as a shear force, used as the weld design value.

References

1. British Standard 4190:1967. Specification for ISO Metric Black Hexagon Bolts Screws and Nuts. British Standards Institution, London (1967).
2. British Standard 3692:1967. Specification for ISO Metric Precision Hexagon Bolts Screws and Nuts. British Standards Institution, London (1967).
3. British Standard 4395:1969. Specification for High Strength Friction Grip Bolts and Associated Nuts and Washers for Structural Engineering: Part 1 General Grade Bolts. Part 2 Higher Grade Bolts. British Standards Institution, London (1967).
4. McGuire, W. *Steel Structures*. Section 5.2. Prentice-Hall International, London (1968).
5. *Steelwork Design*. Guide to BS5950: Part 1:1985. Volume 1. Section Properties. Member Capacities. Constrado, London (1985).
6. Gourd, L.M. *Principles of Welding Technology* (2nd edn.). Edward Arnold, London (1986).

BS5950 clause	Example 5.6 Welded column bracket 5.6.1

305 x 305 x 97 UC

300

2P kN

A B

300

y

x

C

D

weld behind

$\bar{x}$

305

Assume unit weld throat.

$\bar{x} = (2\times305\times305/2)/[(2\times305)+300]$

$= 102$ mm

Area $= 2\times305+300 = 910\,mm^2$

$I_{xx} = (300^3/12)+2\times305\times150^2$

$= \boxed{16\times10^6\,mm^4}$

$I_{yy} = (2\times305^3/12)+300\times102^2$

$+ 2\times305\left((305/2)-102\right)^2 = \boxed{9.4\times10^6\,mm^4}$

$I_{polar} = (16+9.4)\times10^6 = \boxed{25.4\times10^6\,mm^4}$

203 102

$r_1 = (150^2+203^2)^{1/2} = 252$ mm

$r_2 = (150^2+102^2)^{1/2} = 181$ mm

$\alpha = \cos^{-1}(203/252) = 36.3°$

$\beta = \cos^{-1}(102/181) = 55.7°$

F_1 r_1 r_2 α β D C F_2

Vertical shear $= P/910 = 1.1\times10^{-3}P$ kN/mm

BS5950 clause	Example 5.6 Welded column bracket	5.6.2
	$F_1 = P\ (300 - (305/2) + 102) \times 252/25.4 \times 10^6$	
	$= 2.48 \times 10^{-3} P$ kN/mm	
	$F_2 = P\ (300 - (305/2) + 102) \times 181/25.4 \times 10^6$	
	$= 1.78 \times 10^{-3}\ P$ kN/mm	
	By inspection the vector sum of the vertical shear and F_2 is critical	
	$R = P\left\{(1.1 \times 10^{-3})^2 + (1.78 \times 10^{-3})^2 + 2 \times 1.1 \times 1.78 \times 10^{-6} \cos 55.7°\right\}^{1/2}$	
	$= 2.57 \times 10^{-3} P$ kN/mm	
6.6.5.3	For weld leg length s throat size a = 0.7s	
6.6.5.1 Table 36	Grade 43 steel, $p_w = 215$ N/mm² = 0.215 kN/mm²	

weld leg s mm	throat size a mm	P kN
6	4.2	$(0.215 \times 4.2)/2.57 \times 10^{-3} = 351$
8	5.6	$(0.215 \times 5.6)/2.57 \times 10^{-3} = 468$
10	7.0	$(0.215 \times 7.0)/2.57 \times 10^{-3} = 586$

6.6.5.2 [weld assumed to be continued round corners]

BS5950 clause	Example 5.7 Welded beam to column connection 5.7.1
	190 kN; 30 kNm; 140; 386; 300; x x
6.6.5.2	Dimensions are effective lengths
6.6.2.1	Assume unit weld throat
	Area = 2×140 + 2×300
	= 880 mm²
	I_{xx} = 2×140×(386/2)² = 10.4×10⁶ mm⁴
	2×300³/12 = 4.5×10⁶ mm⁴
	14.9×10⁶ mm⁴
	Z for extreme fibre = 14.9×10⁶/(386/2)
	= 77.4×10³ mm³
	Direct stress 190/880 = 0.216 kN/mm
	Bending stress 30×10³/77.4×10³ = 0.388 kN/mm
6.6.5.5	Vector sum = (0.216² + 0.388²)^½ = 0.444 kN/mm
	Weld size s = 0.444/0.215×0.7 = 2.95 mm
	say 3 mm weld (grade 43)

6 Design of element assemblies

6.1 General

In the preceding chapters attention has been focused on the design of individual elements, beams, columns, ties and methods of connecting them together. In practice individual elements are assembled for structural use into frames of various kinds; in this chapter we examine the design of some simple combinations of components. The rules of component design still have to be observed but there is now the added complication of determining the way in which load is shared by the individual members and the interaction between them at the joints in the structure.

In skeletal frames formed of assemblies of members there is usually a clear distinction between continuous structures, in which the joints between the members are themselves rigid, so preserving the relative alignment of members whatever the loading pattern, and pin jointed frames in which the joints allow the members to rotate relative to each other. Examples of the design of both types of frame follow.

6.2 Lattice girders

It was apparent to early designers that economy in beam design would result if a solid web plate were replaced by a series of discrete members, the whole forming with the top and bottom chords a lattice or truss system. The chords need not be parallel; the upper chord may slope to form a pitched roof or north light truss as shown in Figure 6.1.

Depending on the loading and dimensions of the girder the members may be formed from a variety of rolled sections ranging from angles to beams or from built-up sections of varying complexity. Some sections are shown in Figure 6.2. For light trusses single or double angles are often used but there are complications in their design caused by the fact that they may be connected unsymmetrically to their fellow members. Attention must also be paid to the interconnection of the two angles forming double angle members.

The analysis of lattice girders is generally made on the assumption that all joints are pinned and that external forces are applied only at joints. Member forces can then be found by resolution or a graphical method using Bow's

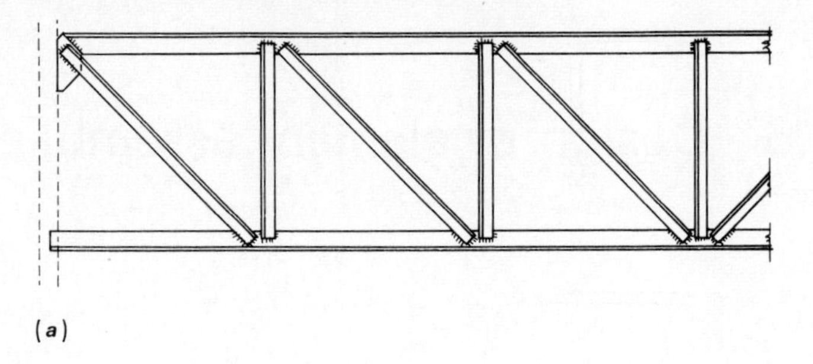

(a)

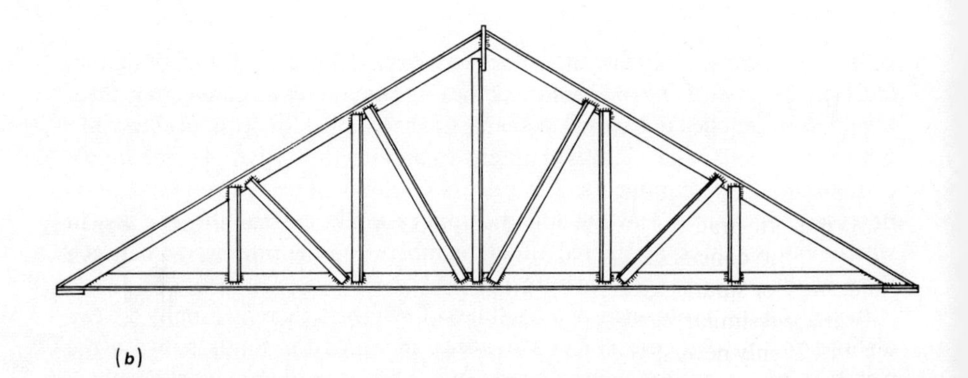

(b)

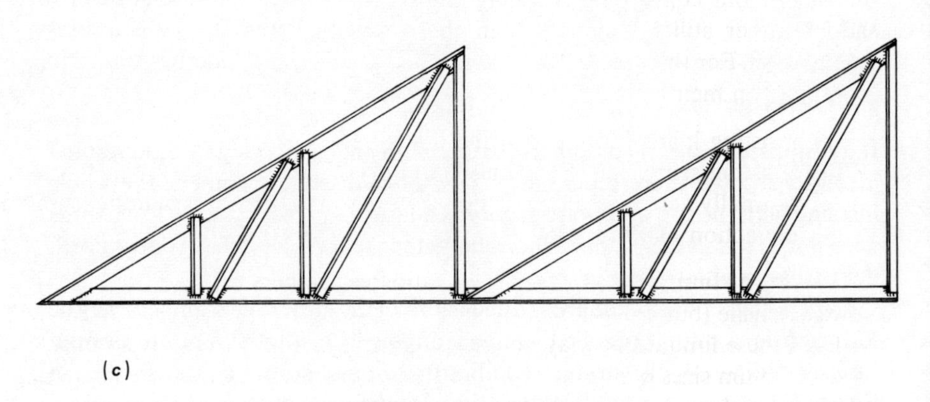

(c)

Figure 6.1 Lattice structures

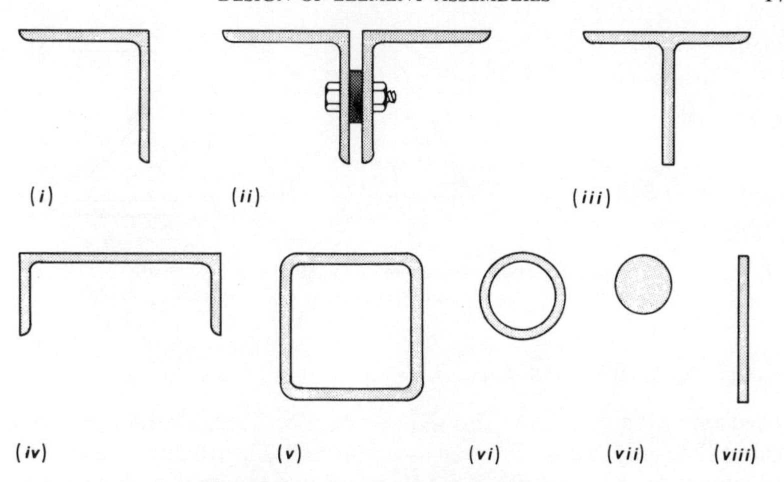

Figure 6.2 Chords of lattice girders:
(i)–(vi) compression, (vii)–(viii) tension members

notation. In practice loads often occur between joints and it may then be necessary to consider local bending in addition to axial force in the members affected. For smaller girders it is usually more economic not to vary the chord section and similarly to keep the diagonals unchanged over the span. In these cases it is only necessary to find the maximum chord forces and the maximum tensile and compressive forces in the diagonals.

While very light members may be desirable for economy there are practical considerations concerning handling and erection which make it prudent to adopt rather stiffer members than those strictly necessary from a stress calculation. For this reason BS 5950 specifies maximum slenderness ratios for compression members:

(a) resisting loads other than wind loads 180
(b) resisting self weight and wind load only 250
(c) normally acting as a tie but subject to reversal of stress resulting from the action of wind 360

There is no limitation on slenderness ratio for members in which the force is always tensile (but see below).

Even these limitations may not be sufficient; Figure 6.3 shows recommended minimum sizes of angles for a roof truss. Despite the lack of size limit on tension members it is often the case that light fittings and other items may be hung from the tension chords of roof trusses.

Extremely light roof structures may also be susceptible to damage from wind action because the dead weight of the roof may well be considerably less than the value of upward force produced by the wind. Not only will the designer need to be aware of reversals of stress in the truss members, he will also need to take care to anchor the truss firmly to its substructure.

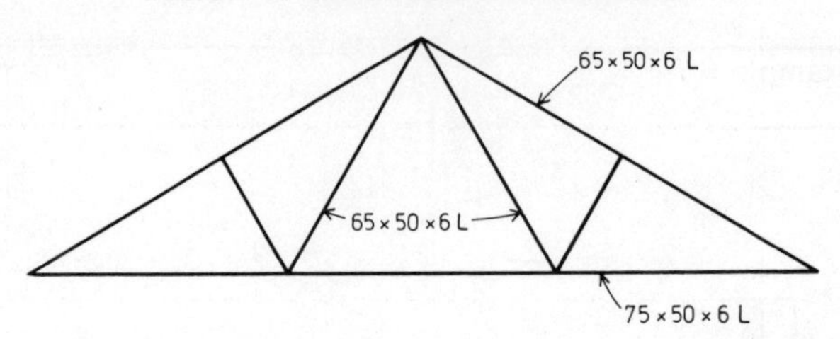

Figure 6.3 Suggested minimum member sizes for trusses
(Main struts 65 × 50 × 6L; main ties 75 × 50 × 6L; others 50 × 50 × 6L)

Example 6.1 Lattice roof beam

The example which follows shows the design of a lattice girder to support a lightweight roof structure over an assembly hall. The column spacing of 1.8 m is dictated by the necessity to conform to a grid imposed by a structural system; if there were not this constraint a greater spacing would be preferable.

Among practical constraints (as opposed to stress limitations) which the designer had to bear in mind for this scheme were the following:

(a) Provision has to be made of a sufficiently wide top chord to the girder to allow the roof panels to be adequately fixed. Lightweight roofs must be anchored to the supporting steelwork against wind suction forces, which are high on flat roofs.
(b) The girder should be sufficiently stiff to withstand transverse forces during handling for erection.
(c) Lateral instability under the dead load of the girder must be avoided before the roof deck is fixed to it.

Member forces are found:

(a) for the chords by dividing the maximum bending moment (assuming the load to be uniformly distributed) by the distance between chord centroids;
(b) for the diagonals by resolving the end reaction into the first two diagonals (one tension, one compression).

Additional stress caused by local bending in the top chord is calculated by considering the chord as a continuous beam supported by the diagonals.

Overall lateral stability of the chord is provided by a positive connection to the roof decking.

6.3 Portal frames

Industrial buildings, for storage or workshop purposes, are very commonly single storey steel frames with sloping rafters of the type shown in Figure 6.4. An economical method of providing this cross-section is to adopt a portal

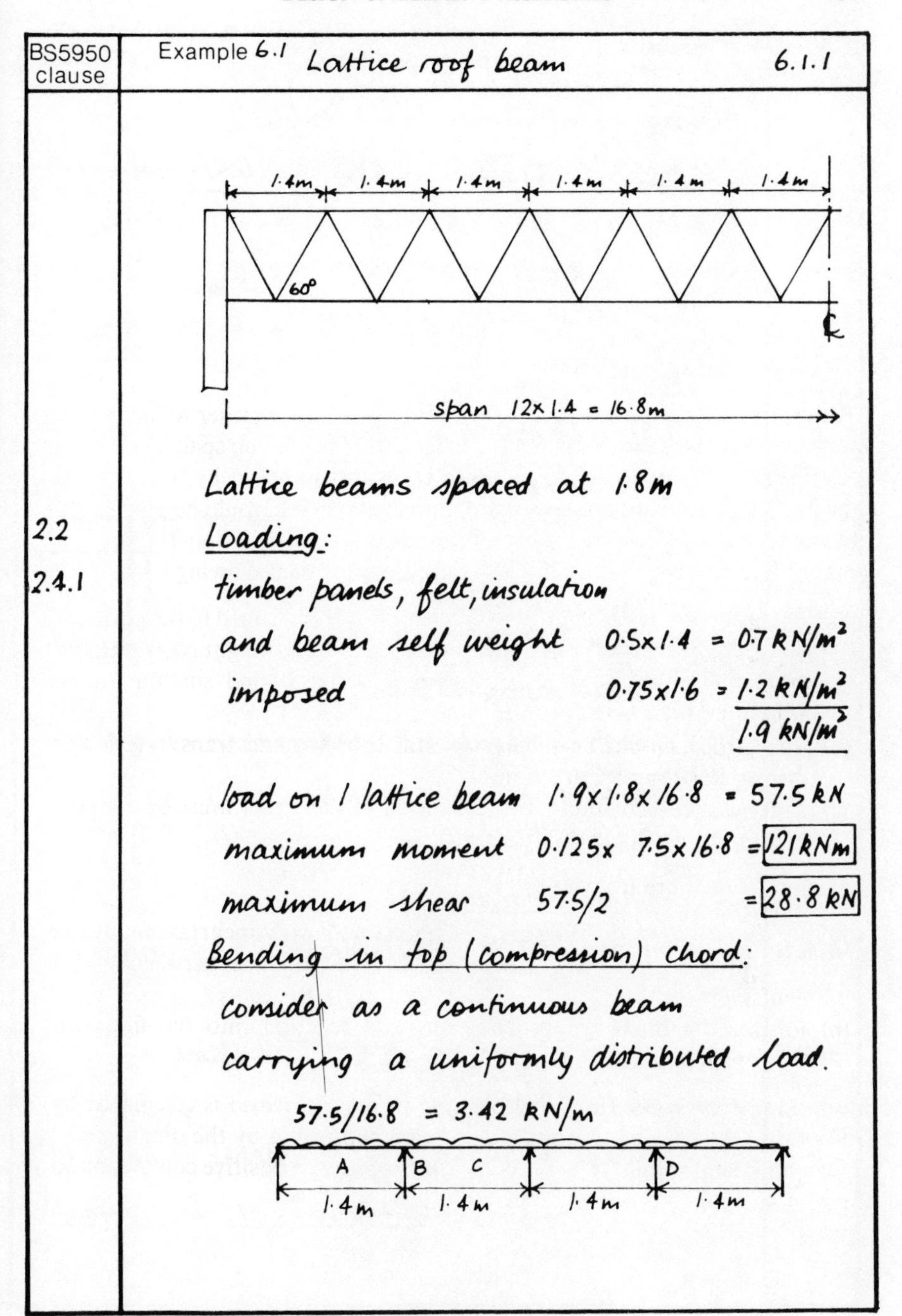

BS5950 clause	Example 6.1 Lattice roof beam	6.1.1

Lattice beams spaced at 1·8m

2.2 — Loading:

2.4.1 — timber panels, felt, insulation and beam self weight $0{\cdot}5 \times 1{\cdot}4 = 0{\cdot}7$ kN/m²

imposed $0{\cdot}75 \times 1{\cdot}6 = 1{\cdot}2$ kN/m²

$1{\cdot}9$ kN/m²

load on 1 lattice beam $1{\cdot}9 \times 1{\cdot}8 \times 16{\cdot}8 = 57{\cdot}5$ kN

maximum moment $0{\cdot}125 \times 7{\cdot}5 \times 16{\cdot}8 =$ **121 kNm**

maximum shear $57{\cdot}5/2 =$ **28·8 kN**

Bending in top (compression) chord:

consider as a continuous beam carrying a uniformly distributed load.

$57{\cdot}5/16{\cdot}8 = 3{\cdot}42$ kN/m

BS5950 clause	Example 6.1 Lattice roof beam 6.1.2
	From continuous beam tables
	$M_A = 0.077 \times 3.42 \times 1.4^2 = 0.516$ kNm
	$M_B = -0.107 \times 3.42 \times 1.4^2 = -0.717$ kNm
	$M_C = 0.036 \times 3.42 \times 1.4^2 = 0.241$ kNm
	$M_D = -0.080 \times 3.42 \times 1.4^2 = -0.536$ kNm
	Chords
	Try 2/65×50×6 unequal angles, long legs connected to gusset
	20.4 x x 8
	$r_{xx} = 2.03$ cm, $r_{yy} = 2.23$ cm
	$Z_{xx} = 12.2$ cm^3, $r_{vv} = 1.06$ cm
	total area 13.2 cm^2
	Force in chord is bending moment/lever arm
	lever arm = 1.4 tan 60/2 = 1.21 m
	force = 121/1.21 = **100 kN**
	Classification
Table 6	grade 43, $p_y = 275$ N/mm^2 ($t < 16$ mm)
	$\varepsilon = 1.0$
Table 7	$b/t = 50/6 < 8.5$ compact
	$d/t = 65/6 < 15$ semi-compact
	$(b+d)/t = (50+65)/6 < 23$ semi-compact
	section is semi-compact

BS5950 clause	Example 6.1 Lattice roof beam 6.1.3
4.7.13	<u>Design as back-to-back strut</u>
4.7.13.1(e)	Interconnect angles, between gussets, to form at least three bays.
4.7.9(c)	check slenderness of angle component ≯ 50
	$L_E = 1400/3 = 467\,mm$
	$L_E/r = \lambda_c = 467/10{\cdot}6 = 44{\cdot}1 < 50$
4.7.9(c)	check slenderness about axis perpendicular to connected surfaces (x-x)
	$\lambda_m = 1400/22{\cdot}3 = 62{\cdot}8$
	$\lambda_b = (62{\cdot}8^2 + 44{\cdot}1^2)^{1/2} = 76{\cdot}7$
4.7.9(c)	check slenderness about axis parallel to connected surfaces (y-y)
	$\lambda = 1{\cdot}4 \times 44{\cdot}1 = 61{\cdot}7$
	∴ design slenderness is $\boxed{76{\cdot}7}$
Table27(c)	$p_c = 168\ N/mm^2$
4.8.3.2(a)	<u>Local capacity</u>
	conservative case $F = 100\,kN$, $M_x = 0{\cdot}536\,kNm$
	$F/A_g p_y = 100 \times 10^3/13{\cdot}2 \times 10^2 \times 275 = 0{\cdot}276$
	$M_x/M_{cx} = 0{\cdot}536 \times 10^6/122 \times 10^3 \times 275 = 0{\cdot}160$
	$0{\cdot}436 < 1{\cdot}0$
4.8.3.3	<u>Overall buckling</u>
4.8.3.3.1	calculate M_b:
4.3.8	use rules for single angles:

BS5950 clause	Example 6.1 Lattice roof beam 6.1.4
4.3.8	L/r_{yy} = 1400/22·3 = 62·8 < 100
	M_b = 0·8 × 275 × 12·2 × 10^{-3} = 2·68 kN m
	conservative case F = 117 kN, M_x = 0·536 kNm, m = 1
	$F/A_g p_c$ = 100 × 10^3/13·2 × 10^2 × 168 = 0·451
	$m M_x/M_b$ = 1·0 × 0·536/2·68 = 0·200
	0·651 < 1·0
	Adopt 2/65 × 50 × 6 angles for chords
	Diagonals
	tensile force P = R/sin 60°
	P = 28·8/sin 60° = **33·3 kN**
	compressive force Q = **33·3 kN**
	A, C, P, Q, 1210, 60°, 60°, B, R
	Strut B-C
4.7.10	Try 50 × 50 × 5 equal angle
	length = 1210/sin 60° = 1400 mm
	r_{aa} = 1·51 cm, r_{vv} = 0·973 cm, A = 4·2 cm^2
4.7.10.2(a) Table 28	**slenderness:**
	$0·85 L/r_{vv}$ = 0·85 × 1400/0·973 × 10 = 122
	$0·7 L/r_{aa} + 30$ = (0·7 × 1400/1·51 × 10) + 30 = 94·9
Table 27c	λ = 122, p_c = 95 N/mm^2
	P_c = 4·2 × 95 × 10^{-1} = **39·9 kN** > 33·3 kN
	Adopt 50 × 50 × 5 angles for diagonal struts

BS5950 clause	Example 6.1 Lattice roof beam 6.1.5
	Tie A-B
	Try 50x50x5 equal angle connected through one leg
4.6.3.1	net sectional area of connected leg (welded)
	$a_1 = 4.8/2 = 2.4\,cm^2 \doteq a_2$
	$3a_1/(3a_1 + a_2) = 0.75$
	$A_e = 2.4 + 0.75 \times 2.4 = 4.2\,cm^2$
4.6.1	$P_t = 4.2 \times 275 \times 10^{-1} = \boxed{115\,kN} > 33.3\,kN$
	Adopt 50x50x5 angles for all diagonals

BS5950 clause	Example 6.2 Portal frame 6.2.1
	$h_2 = 6.0\text{m}$, $h_1 = 5.5\text{m}$, 16.2m, $L = 30.0\text{m}$
	Frame spacing 6.0m
	Steel grade 50
	Loading
	dead (on slope)
	sheeting and insulation 0.21 kN/m²
	purlins 0.09
	rafters 0.10
	0.40 kN/m²
	plan equivalent $0.4 \times 16.2/15$ = **0.432 kN/m²**
	imposed (on plan) **0.750 kN/m²**
	wind - considered not critical
5.3 5.5.3	a. Plastic design.
	For a uniformly distributed roof load intensity w:
	$M_p = \frac{wL^2}{8} \times \frac{1}{1+\frac{k}{2} + (1+k)^{1/2}}$
	where $k = \frac{h_2}{h_1} = \frac{6.0}{5.5} = 1.09$

BS5950 clause	Example 6.2 Portal frame 6.2.2
	Factored loading on rafter:
2.4	dead $0.432 \times 1.4 \times 6.0 = 3.63$ kN/m
Table 2	imposed $0.75 \times 1.6 \times 6.0 = 7.2$ kN/m
	10.8 kN/m
	$M_p = \frac{10.8 \times 30^2}{8} \times \frac{1}{1 + \frac{1.09}{2} + (1 + 1.09)^{1/2}} = \boxed{406 \text{ kNm}}$
	Factored axial load in column
	$F = 10.8 \times 30/2 = \boxed{162 \text{ kN}}$
	Selection of a suitable section is made using Constrado Guide, volume 1:
	Try 457 × 152 × 60 UB, grade 50
	moment capacity $M_{cx} = \boxed{454 \text{kNm}} > 406$ (p295)
	check interaction between compression and bending
	squash load $P_z = 2690$ kN
	$F/P_z = 162/2690 = 0.06$
	$M_{rx} = \boxed{452 \text{kNm}} > 406$ kNm (p295)
	Section is suitable from consideration of moment capacity and axial load.
	Further checks, outside the scope of this example, will be necessary. Refer to BS 5950 section 5.3 Plastic Design and section 5.5.3 Portal Frames, Plastic Design.

BS5950 clause	Example 6.2 Portal frame 6.2.3
5.2 5.5.2	b. Elastic design Elastic analysis is carried out under factored loads. The distribution of bending moment may be found by a variety of hand methods or by computer.

Critical load condition is at B or D

M_B = 524 kN m

Axial load (as for plastic design) 162 kN

From Constrado Guide, volume 1

Try 457 x 152 x 74 UB (p 294)

M_{cx} = 551 kNm, M_{rx} = 549 kNm > 524

Further checks, outside the scope of this example, will be necessary. Refer to BS5950 section 5.5.2, Portal Frames, Elastic Design.

If the sections in a. and b. are adopted the saving in section weight by plastic design = (74-60)/60 = 23%.

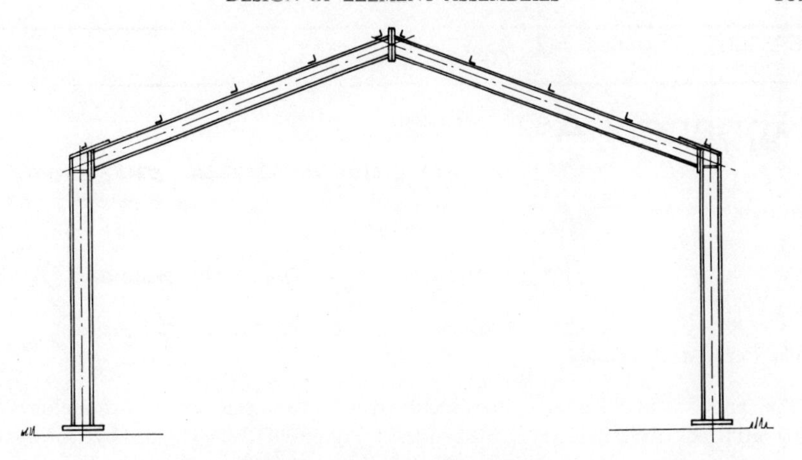

Figure 6.4 Pitched portal frame

frame composed of universal beam columns and rafters. Site connections, if required, are made using high-strength friction grip bolts; shop connections are welded. Erection of these portal frames can be very rapid and the addition of ribbed steel roof and wall cladding provides a weatherproof building in a very short time.

The example covers only the outline of calculations for the main frame members. Such detailed considerations as bracing, foundations and so on are beyond the scope of the example and for these reference should be made to a specialist text.

It should be noted that plastic design shows economy in the weight of main members but that it will be necessary to check stability and deflection (1).

Reference

1. Horne, M.R. and Morris, L.J. *Plastic Design of Low-rise Frames.* Granada, London (1985).

Appendix A

The Perry strut formula

A pin ended strut of length L, cross-sectional area A and elastic section modulus Z has an initial lack of straightness $v_0 = \Delta sin(\pi z/L)$ (Figure A1). When loaded by a force P the compressive stress at any cross section is the sum of direct compression P/A and bending compression M_z/Z.

If the additional deflection at z from the origin is v then

$$M_z = P(v + v_0) = -EI\frac{d^2v}{dz^2}$$

which gives

$$\frac{d^2v}{dz^2}k^2v + k^2\Delta\sin\frac{\pi z}{L} = 0 \quad \text{(A1)}$$

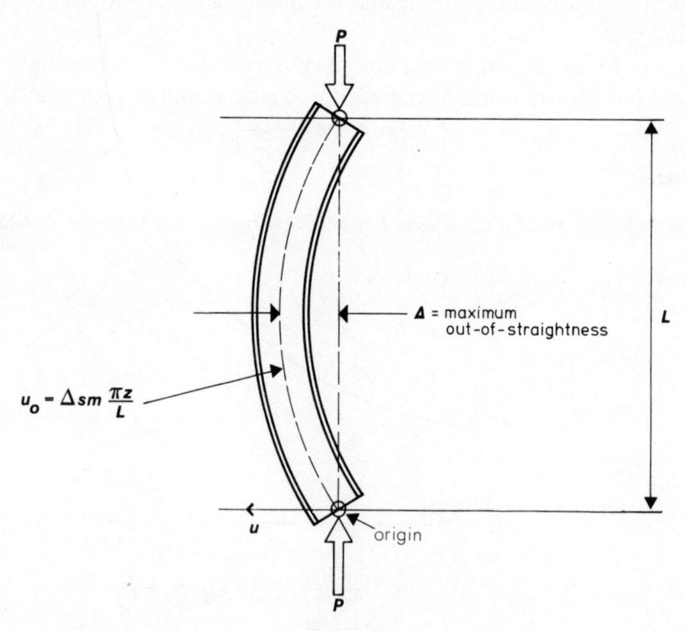

Figure A1

where

$$k^2 = P/EI$$

Equation (A1) has the general solution

$$v = A \sin kz + B \cos kz + \frac{\Delta}{((\pi^2/k^2L^2) - 1)} \sin \frac{\pi z}{L}$$

The boundary conditions are:

$$\left.\begin{matrix} z=0 \\ v=0 \end{matrix}\right\} \quad B=0 \qquad \left.\begin{matrix} z=L \\ v=0 \end{matrix}\right\} \quad A=0$$

Now $P_E = \pi^2 EI/L^2$, and $k^2 = P/EI$. Hence $P_E/P = \pi^2/k^2L^2$ so that

$$v = \frac{\Delta}{((P_E/P) - 1)} \sin \frac{\pi z}{L}$$

and the total deflection $(v + v_0)$ is

$$\left[\left(\frac{1}{\dfrac{P_E}{P} - 1}\right) + 1\right] \Delta \sin \frac{\pi z}{L}$$

$$= \frac{1}{\left(1 - \dfrac{P}{P_E}\right)} \Delta \sin \frac{\pi z}{L}$$

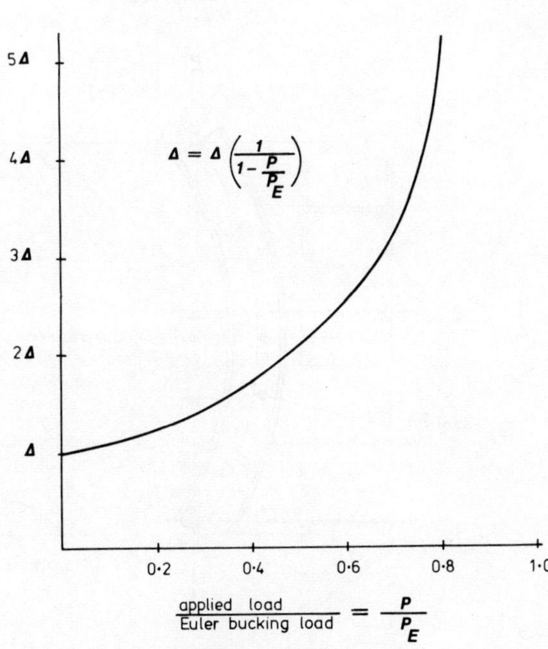

Figure A2

Thus the maximum deflection at mid-height is

$$\left[\frac{1}{\left(1-\frac{P}{P_E}\right)}\right]\Delta$$

The bracketed term is known as the amplification factor, the effect of which is illustrated in Figure A2.

The total compressive stress is

$$\frac{P}{A}+\frac{P}{Z}\Delta\left[\frac{1}{\left(1-\frac{P}{P_E}\right)}\right]$$

If $\frac{P}{A}=p_c, \frac{P_E}{A}=p_E$

and the limiting total stress is p_y

$$p_c\frac{A\Delta}{Z}\left[\frac{1}{\left(1-\frac{p_c}{p_E}\right)}\right]=p_y$$

The expression $A\Delta/Z=y\Delta/r^2$ is termed the imperfection factor η.

$$p_c\eta\left(\frac{p_E}{p_E-p_c}\right)=p_y-p_c$$

or

$$(p_E-p_c)(p_y-p_c)=\eta p_E p_c$$

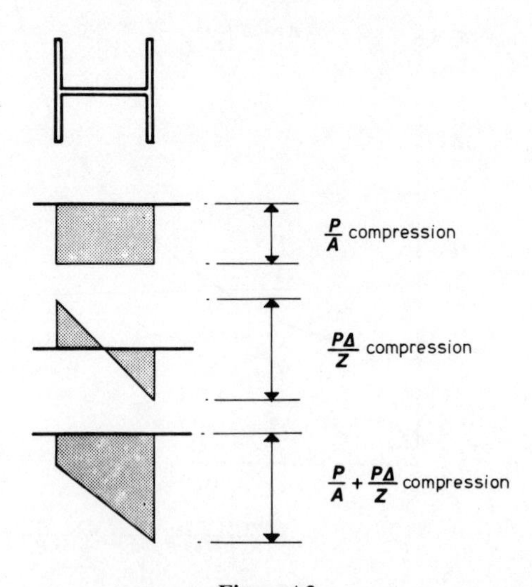

Figure A3

Bibliography

Ashby M.F. and Jones, D.R.H. *Engineering Materials. An Introduction to their Properties and Applications.* Pergamon, Oxford (1981).

Ballio, G. and Mazzolani, F.M. *Theory and Design of Steel Structures.* Chapman and Hall, London (1983).

Billington, David P. *The Tower and the Bridge. The New Art of Structural Engineering.* Basic Books Inc., New York (1983).

Collins, A.R. *Structural Engineering – Two Centuries of British Achievement.* Tarot Print Ltd., Chislehurst (1983).

Fisher, J.W. and Struik, J.H.A. *Guide to Design Criteria for Bolted and Riveted Joints.* John Wiley, New York (1974).

Higgins, R.A. *Engineering Metallurgy.* 1. *Applied Physical Metallurgy* (5th edn.). Hodder and Stoughton, London (1983).

Horne, M.R. *Plastic Theory of Structures* (2nd edn.). Nelson, London (1979).

Johnson, R.P. *Composite Structures of Steel and Concrete.* Vol. 1: *Beam, Columns, Frames and Applications in Building.* Granada, London (1975).

Johnston, B.G. (ed.) *Column Research Council Guide to Stability Design Criteria* (3rd edn.). John Wiley, New York, 1976.

McGuire, W. *Steel Structures.* Prentice-Hall, Englewood Cliffs (1972).

Nethercot, D.A. *Limit States Design of Structural Steelwork.* Van Nostrand Reinhold, Wokingham (1986).

Stephens, J.H. *The Guinness Book of Structures.* Guinness Superlatives, Enfield (1976).

Trahair, N.S. *The Behaviour and Design of Steel Structures.* Chapman and Hall, London (1977).

Index